A-LEVEL YEAR 2
STUDENT GUIDE

CR

Biology A

Module 5

Communica   , homeostasis and energy

Richard Fosbery

Philip Allan, an imprint of Hodder Education, an Hachette UK company, Blenheim Court, George Street, Banbury, Oxfordshire OX16 5BH

Orders

Bookpoint Ltd, 130 Park Drive, Milton Park, Abingdon, Oxfordshire OX14 4SE

tel: 01235 827827

fax: 01235 400401

e-mail: education@bookpoint.co.uk

Lines are open 9.00 a.m.–5.00 p.m., Monday to Saturday, with a 24-hour message answering service. You can also order through the Hodder Education website: www.hoddereducation.co.uk

© Richard Fosbery 2016

ISBN 978-1-4718-5915-1

First printed 2016

Impression number 5 4 3 2 1

Year 2020 2019 2018 2017 2016

This guide has been written specifically to support students preparing for the OCR A A-level Biology examinations. The content has been neither approved nor endorsed by OCR and remains the sole responsibility of the author.

Cover photo: Argonautis/Fotolia; p. 96/Science Photo Library; other photographs Richard Fosbery

Typeset by Integra Software Services Pvt. Ltd, Pondicherry, India

Printed in Italy

Hachette UK's policy is to use papers that are natural, renewable and recyclable products and made from wood grown in sustainable forests. The logging and manufacturing processes are expected to conform to the environmental regulations of the country of origin.

Contents

Content Guidance

Questions & Answers

■ Getting the most from this book

Exam-style questions

Commentary on the questions

Tips on what you need to do to gain full marks, indicated by the icon **e**

Sample student answers

Practise the questions, then look at the student answers that follow.

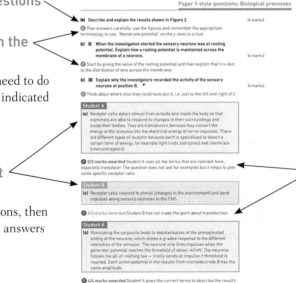

Commentary on sample student answers

Find out how many marks each answer would be awarded in the exam and then read the comments (preceded by the icon **e**) following each student answer showing exactly how and where marks are gained or lost.

■ About this book

This guide is the third in a series of four covering the OCR A-level Biology A specification. It covers Module 5: Communication, homeostasis and energy, and is divided into two sections:

- The **Content Guidance** provides key facts and key concepts, and links with other parts of the A-level course. The synoptic links should help to show you how information in this module is useful preparation for other modules.
- The **Questions & Answers** section contains two sets of questions, giving examples of the types of question to be set in the A-level papers. The first of these — Paper 1: Biological processes — has some multiple-choice questions and some structured questions, together with answers to all the structured questions written by two students. There are examiner comments on all the answers. The second paper — Paper 3: Unified biology — has some questions that set the topics in Module 5 in a wider context.

This guide is not just a revision aid. You will gain a much better understanding of the topics in Modules 2 to 6 if you read around the subject. I have suggested a few websites that you can use for extra information. These will help especially with topics that are best understood by watching animations of processes taking place. As you read this guide remember to add information to your class notes.

The Content Guidance section will help you to:

- organise your notes and to check that you have highlighted the important points (key facts) — little 'chunks' of knowledge that you can remember
- understand how these 'chunks' fit into the wider picture. This will help to support:
  - Module 6, which is covered in the fourth student guide in this series
  - Modules 2, 3 and 4, which you are most likely to have studied in the first year of your course; these are covered in the first two student guides in this series
- check that you understand the links to the practical work, since you must expect questions on practical work in your examination papers. Module 1 lists the details of the practical skills you need to use in the papers
- understand and practise some of the maths skills that will be tested in the examination papers — look out for this icon for examples: 🖩

The **Questions & Answers** section will help you to:

- understand which examination papers you will take
- check the way examiners ask questions in the A-level papers
- understand what examiners mean by terms like 'explain' and 'describe'
- interpret the question material — especially any data that the examiners give you
- write concisely and answer the questions that the examiners set

Content Guidance

Communication and homeostasis

■ Communication and homeostasis

Key concepts you must understand

If they are to survive, all organisms must detect stimuli (changes) in their surroundings and in their internal environment, and respond to them. Complex, multicellular organisms are composed of specialised cells, tissues, organs and organ systems. Cells communicate with each other by secreting chemicals that diffuse into the immediate surroundings and are detected by adjacent cells. However, this method is only effective over short distances. So that cells throughout the body can function efficiently and effectively, organisms need long-distance communication systems to coordinate and synchronise the activities of their cells, tissues and organs. Animals have two such systems:

- the nervous system
- the endocrine (hormonal) system

Four ways in which cells signal to one another are:

- paracrine secretion, over short distances
- nerve cells using electrical impulses to cover long distances, but releasing chemicals where they terminate
- endocrine secretion
- neurosecretion

In endocrine secretion and neurosecretion, chemicals are released into the blood to travel long distances. These four methods of signalling are shown in Figure 1.

To function efficiently organisms have control systems to keep internal conditions near constant, a feature known as homeostasis. This requires information about conditions inside the body and in the surroundings. These conditions are detected by sensory cells. Some of the physiological factors controlled in homeostasis in mammals are:

- core body temperature
- blood glucose concentration
- concentration of ions in the blood, for example Na^+, K^+ and Ca^{2+}
- water potential of the blood

Exam tip

As you look at Figure 1 remember what you learned about cell signalling and cell surface receptors in Module 2.

Exam tip

Histamine, antidiuretic hormone, glucagon and acetylcholine are examples of cell signalling molecules. There are many others, especially in the central nervous system.

Homeostasis
Maintenance of physiological factors of the body within narrow limits.

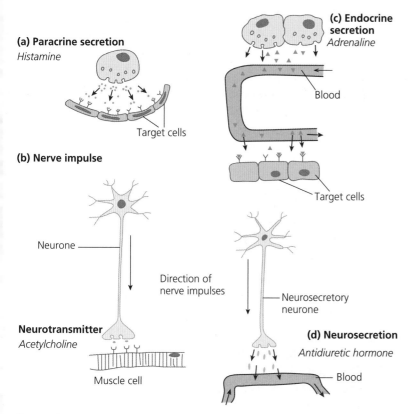

(a) Paracrine secretion
Histamine

Target cells

(b) Nerve impulse

Neurone

Direction of
nerve impulses

Neurotransmitter
Acetylcholine

Muscle cell

**(c) Endocrine
secretion**
Adrenaline

Blood

Target cells

Neurosecretory
neurone

(d) Neurosecretion

Antidiuretic hormone

Blood

Figure 1 Cell signalling involves the release of chemicals and their detection

Key facts you must know
Homeostasis and thermoregulation

Homeostasis is the maintenance of near constant internal conditions. It involves detecting changes inside and outside the body and making responses that counteract changes in the internal environment.

Birds and mammals maintain a near constant core body temperature independent of the temperature of their surroundings. They are **endotherms** — they generate heat and conserve it in their bodies if it is cold, and have mechanisms to lose heat if they are hot. Other animals and plants are unable to do this and are known as **ectotherms**. Their main source of body heat is from their surroundings, and their body temperatures fluctuate with the ambient temperature. It is a mistake to call ectotherms, such as lizards, snakes and fish, 'cold-blooded'. On a hot day a lizard is hot-blooded and tropical fish are definitely warm-blooded. Although ectotherms are dependent on the temperature of their surroundings, many organisms are able to maintain a different temperature from that of their surroundings.

- Some lizards are able to regulate their temperature using behavioural methods such as basking in the sunshine and moving into the shade.
- Desert plants are often a few degrees cooler than the air temperature on hot days.
- Arctic and Antarctic plants and insects are often dark to absorb heat so that their temperatures are several degrees higher than air temperature.

Exam tip

The ambient temperature is the temperature of an organism's surroundings. This may be air or water depending on the species. A key principle of temperature control is detecting a decrease in the ambient temperature and taking actions to reduce heat loss before the core body temperature falls.

Knowledge check 1

Distinguish between ectotherms and endotherms.

The normal range of human body temperature is 36.0–37.6°C. Most birds have higher temperatures, for example 40–42°C.

To regulate body temperature mammals have:
- **receptors** that detect changes in the temperature of the surroundings and in the internal (core) body temperature — these are **peripheral receptors**
- a control centre that receives information about changes in temperature and sends instructions to the body to make adjustments to counteract changes in body temperature
- **effectors** in the skin and muscles that produce heat, promote heat loss or promote heat conservation

The thermoregulatory centre is situated in the **hypothalamus** in the brain. It lies just above the pituitary gland in the centre of the head. The anterior (front) part of the hypothalamus responds to increases in temperature by promoting heat loss. The posterior (back) part responds to decreases in temperature by conserving heat and stimulating heat production.

The hypothalamus constantly compares information about the temperature of the surroundings and the internal temperature with the **set point**, and acts to keep the difference between the *actual* body temperature and the set-point temperature as small as possible.

The control centre needs information about whether instructions sent to effectors to carry out corrective actions to produce heat, lose heat or gain heat are effective. This involves feedback in which a change in the factor being controlled acts as a stimulus, which leads to a response that counteracts the change in the factor. The effects of corrective actions are monitored by the hypothalamus and adjustments are made to keep the temperature within narrow limits.

This control method is called **negative feedback** (Figure 2). Negative feedback mechanisms keep a physiological factor, such as core body temperature, within narrow limits.

Receptor Specialised cell or tissue, or a nerve cell ending, that detects a stimulus and causes impulses to travel in neurones to the CNS.

Effector A cell, tissue or organ that carries out a function in response to a signal.

Set point The value of a physiological factor that the body tries to keep constant all the time.

Negative feedback Method of control that makes corrective actions to keep a certain factor (e.g. body temperature) within narrow limits.

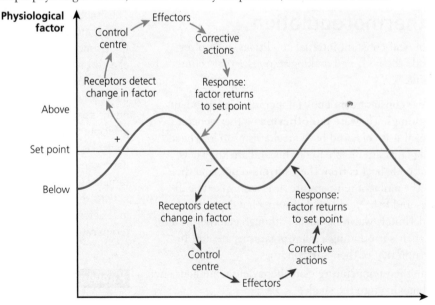

Figure 2 The principle of negative feedback

The flow charts in Figure 3 show how the hypothalamus is involved in controlling body temperature in humans. Part (a) shows what happens when you absorb heat or generate heat internally during exercise. Part (b) shows what happens when the ambient temperature is low and you lose heat to your surroundings. Apply the key terms to each part of the flow chart by writing an explanation of the control of body temperature.

When the temperature of the surroundings increases, the skin absorbs heat. This stimulates peripheral (outer as opposed to central) temperature receptors in the skin. During exercise, heat is released from muscles and increases the blood temperature. This increase is detected by central temperature receptors in the spinal cord and in the hypothalamus. When the body temperature rises for either or both of these reasons, the hypothalamus stimulates the **loss of heat** (Figure 3a). Many mammals have few sweat glands and are covered in fur. In addition to the changes shown, in many mammals the heat loss centre also stimulates:

- erector pili muscles in the skin to relax so that hairs lie flat and less air is trapped by the fur
- shallow breathing (panting), which causes loss of heat by evaporation of water from the mouth and gas exchange system

Animals can also change their behaviour to lose heat. For example they can:

- move into the shade or into a burrow
- become inactive, so they do not produce heat through muscle contraction
- avoid being active during the hottest part of the day
- drink something cold so that heat is transferred from the blood to water in the stomach

When the temperature of the surroundings decreases, temperature receptors in the skin are stimulated. They send information to the hypothalamus — early warning that if nothing is done heat will be lost and the core body temperature will fall. To prevent this happening, the hypothalamus stimulates the **conservation of heat** (Figure 3b). In mammals with fur, the erector pili muscles in the skin contract and raise the hairs so that a thicker layer of air is trapped in the fur and convection currents do not reach the skin surface.

If the blood temperature decreases, the hypothalamus stimulates the **production of heat** in the liver and muscles. Young mammals use brown fat tissue to produce extra heat. Cells in brown fat tissue are full of mitochondria that generate heat rather than ATP. Brown fat is also present in mammals that live in cold climates and hibernate. Changes in behaviour can also be used:

- Animals can move somewhere warm or curl up to reduce the surface area exposed to the air.
- Humans can drink something hot so that heat is transferred from the drink to the blood flowing through the stomach wall.
- Lizards can bask in the sunshine to absorb heat in the mornings.

Exam tip

The key message here is if you get cold, the corrective actions you take attempt to conserve what heat you have in your body before using energy reserves to generate more heat.

Knowledge check 2

Name the thermoregulatory centre in the brain.

Exam tip

When you write answers about homeostasis you should try to use the following key terms: negative feedback, set point, control centre, receptor, effector and corrective actions.

Exam tip

Make a list of the different ways in which heat is transferred and apply them to thermoregulation in animals.

Knowledge check 3

Explain briefly how an endothermic animal, such as a bird or mammal, prevents its core body temperature decreasing to the ambient temperature on a cold day.

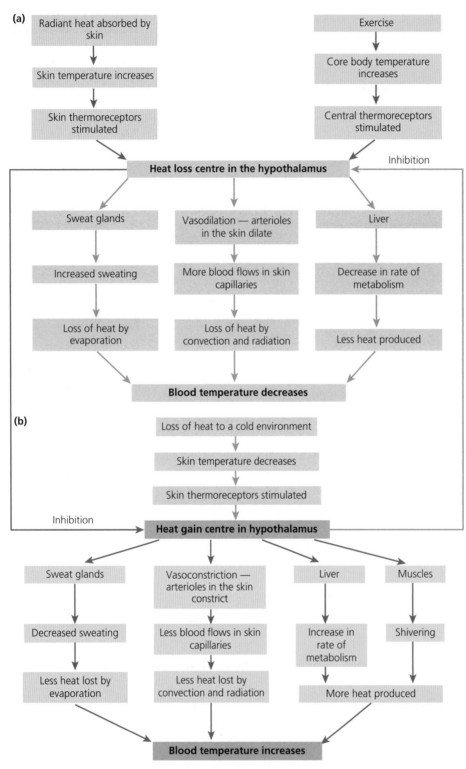

Figure 3 The hypothalamus is the control centre for many aspects of homeostasis, including body temperature. Skin receptors are peripheral thermoreceptors that detect changes in ambient temperature

Britain's smallest mammal, the pygmy shrew, *Sorex minutus*, eats up to 125% of its body mass each day. Facts like this are useful ways to remember biological principles.

Exam tip

Small birds and mammals have large surface area-to-volume ratios, so lose heat quickly. Some mammals and at least one species of bird are unable to maintain a constant body temperature when the external temperature falls, and they hibernate.

The behavioural methods shown by lizards are an example of adaptation. Ectothermic animals are inactive at low temperatures. Endotherms compete well because they can be active all the time. This has helped mammals and birds to colonise cold environments where there is little or no competition from reptiles.

Positive feedback is a control mechanism that detects a difference between the physiological factor and its set point, which leads to effectors *increasing* that difference. Positive feedback is not involved in homeostasis as it does not act to keep anything constant.

An example of positive feedback is what happens when the body's ability to control temperature fails. Thermoregulation stops when the body temperature reaches about 45°C. During heat stroke, the temperature of the body rises as heat is absorbed from the surroundings, but the hypothalamus does not stimulate sweat glands to produce sweat. The increase in temperature stimulates metabolism, which releases more heat and so the temperature increases. This positive feedback is dangerous and can be fatal.

Birth in mammals is controlled by both the nervous and hormonal systems working together. This is another example of positive feedback: an increase in pressure in the uterus stimulates greater muscle contractions that increase the stimulus.

Synoptic links

You may be asked about the advantages and disadvantages of maintaining a constant temperature. You could answer this in terms of enzyme action, using information from Module 2 about the effect of changes in temperature and pH on enzyme activity. Maintenance of near constant conditions means that enzyme activity is stable, as are the metabolic processes (e.g. respiration) that enzymes control. Compared with similar-sized ectotherms, mammals and birds have to eat more food because much of the energy in their food is used to generate heat. Small mammals and birds eat more than their body mass each day.

Positive feedback
Method of control that responds to a change in a factor by making the change even greater.

Exam tip

Positive feedbacks are not sustainable. They usually happen for a short time and then stop, as in action potentials in neurones (p. 30).

Knowledge check 4

Explain the difference between control by negative feedback and control by positive feedback.

Summary

- Multicellular organisms (plants and animals) need communication systems so that they can respond to internal changes and changes in their environment; these systems use cell signalling methods to coordinate the activities of different organs.
- Animal cells communicate by using the nervous system and the endocrine (hormonal) systems.
- Homeostasis is the maintenance of near constant conditions in the body. Core body temperature is an example of a physiological factor that is kept within narrow limits.
- The set point is the value of any physiological factor that the body keeps constant. Negative feedback is the way in which changes in a factor, such as body temperature, stimulate corrective actions to restore the factor to its set point.
- Internal and external changes are detected by receptors that communicate with a central control. These changes are stimuli for responses made by effector cells, tissues and organs that carry out corrective actions.
- Ectotherms rely on external sources of heat. Endotherms generate their own heat and conserve it in their bodies using fat and fur or feathers. Endotherms use physiological and behavioural methods to maintain a constant body temperature. Ectotherms primarily use behavioural methods, for example basking in the sunshine.
- Positive feedback systems respond to a change in a physiological factor by increasing the change. These systems, such as the control of birth, are not homeostatic.

Excretion as an example of homeostatic control

Key concepts you must understand

Excretion is the removal from the body of toxic waste products of metabolism and substances that are in excess of requirements.

Metabolic waste products are:

- carbon dioxide from respiration, which is lost through the gas exchange system
- nitrogenous waste in the form of ammonia, urea and uric acid

Carbon dioxide is produced by decarboxylation of respiratory substrates (p. 67). Ammonia is produced by the deamination of excess amino acids (Figure 8, p. 14). If allowed to accumulate, both carbon dioxide and ammonia would change the pH of cytoplasm and body fluids, and this would cause enzymes to work less efficiently. Urea and uric acid are both harmful if they accumulate.

The excretory system consists of the liver, which produces many of the excretory products, and the kidneys, ureters, bladder and urethra. The lungs remove most of the carbon dioxide, so can also be considered part of the excretory system.

Key facts you must know

The liver

The liver is a large organ situated in the abdomen to the left of the stomach, when viewed from the front. It carries out many functions including as an effector in temperature regulation (p. 9), regulation of metabolism, regulating amino acids, protein synthesis, synthesis of cholesterol and bile salts, digestion, assimilation and excretion.

Figures 4, 5 and 6 show the structure of the liver.

Excretion Removal from the body of toxic waste products of metabolism and substances that are in excess of requirements.

Exam tip

If all the carbon dioxide from respiring tissues were to dissolve in the plasma, the pH would decrease significantly and death would occur in a very short time. In Module 3 you studied the ways in which carbon dioxide is transported in the blood that avoid fluctuations in blood pH.

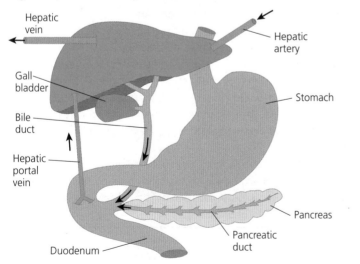

Figure 4 The gross structure of the liver, its blood supply and connection to the duodenum

The liver is divided into sections, known as lobules, which are separated by areas of connective tissue. Oxygenated blood flows into the lobules in branches of the hepatic artery and deoxygenated blood rich in absorbed food from the digestive system enters in branches of the hepatic portal vein. This blood flows through wide capillaries known as sinusoids that are lined by an incomplete layer of endothelial cells, which allows blood to reach the surface of cells, known as hepatocytes, where exchanges of substances occur between blood and cells.

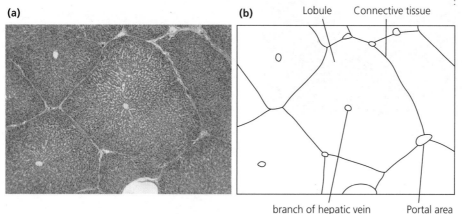

Figure 5 (a) A low-power photomicrograph of some liver lobules. (b) A plan drawing made from the photomicrograph

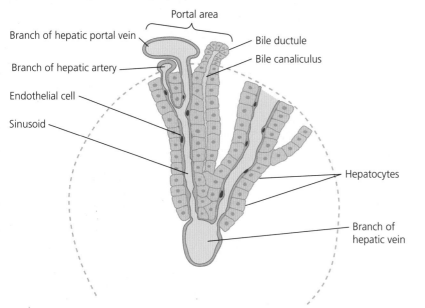

Figure 6 Arrangement of blood vessels, sinusoids, hepatocytes and bile canaliculi in each liver lobule

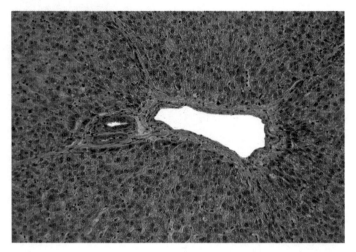

Figure 7 A high-power photomicrograph of a portal area from a liver lobule. Compare with Figure 6 to identify branches of the hepatic artery and hepatic portal vein and a bile canaliculus

Hepatocytes make bile, which is a digestive secretion that is stored in the gall bladder and enters the duodenum. Bile contains bile pigments, which are excretory products made from haemoglobin, and bile salts for the emulsification of fats. These substances enter the little channels (canaliculi), which join together to form bile ducts that drain into the gall bladder and the duodenum. The blood flows along the sinusoids from the branches of the hepatic artery and the hepatic portal vein to drain into a branch of the hepatic vein. This deoxygenated blood flows back to the heart through the vena cava.

Formation of urea

The body requires amino acids to make proteins. At any one time, the intake of amino acids may exceed the requirements. Excess amino acids cannot be stored and are broken down by enzymes in the liver. The amino group from each amino acid is removed to form ammonia (NH_3) and an organic acid in **deamination** (Figure 8).

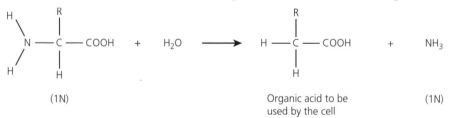

Figure 8 Deamination. The amino group ($-NH_2$) is removed from amino acids to form ammonia. Most deamination occurs in the matrix of mitochondria to prevent ammonia damaging the rest of the cell

The organic acid is used in respiration or in the synthesis of other compounds. Ammonia is made less harmful by the reactions of the **ornithine** (or **urea**) **cycle**. Figure 9 shows an outline of the ornithine cycle (reactions 2 to 5). Reactions 1 and 2 occur in the matrix of mitochondria (Figure 47) and the others occur outside mitochondria in the cytosol.

> **Exam tip**
>
> The largest vessel in Figure 7 is a branch of the hepatic portal vein. To its left is a bile canaliculus and below it a branch of the hepatic artery.

> **Exam tip**
>
> The ornithine cycle results in the production of one molecule of urea from two amino groups and one molecule of carbon dioxide. The advantage of a cycle is that only small quantities of the intermediates are needed to process large quantities of amino groups and carbon dioxide.

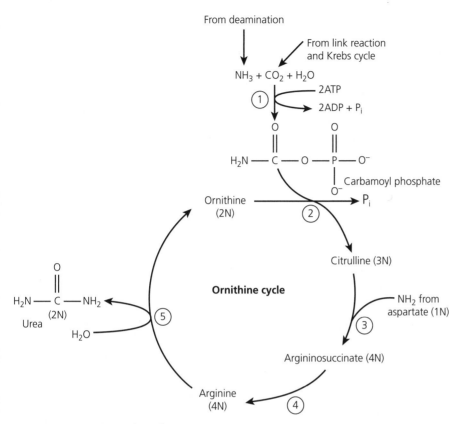

Figure 9 The ornithine (urea) cycle. The numbers of nitrogen atoms in each compound are shown in brackets. Aspartate is an amino acid that is formed from others as a way to remove amino groups

Urea is soluble in water and is less toxic than ammonia. It diffuses readily through membranes and so leaves the hepatocytes and is transported to the kidneys dissolved in the blood plasma. We also excrete tiny quantities of ammonia and uric acid, which is made from purines.

Detoxification

The liver breaks down toxic substances, such as alcohol. This is known as **detoxification**.

The metabolism of alcohol is linked with the reduction of NAD, which is recycled by oxidation in mitochondria. This in turn generates ATP for the cell. Since the metabolism of alcohol generates plenty of energy as ATP, liver cells do not use as much fat as usual, so this is stored within the cells, giving rise to the condition 'fatty liver'. The fat stored in the liver reduces the efficiency of the hepatocytes in carrying out their many functions. Fatty liver can lead to life-threatening conditions such as cirrhosis, which is on the increase in the UK among young people who misuse alcohol.

Other drugs, such as paracetamol, steroids and antibiotics, are also broken down in the liver. Hepatocytes contain catalase, the enzyme that breaks down hydrogen peroxide, which is a toxic end product of metabolism.

Exam tip

Make a labelled diagram of a mitochondrion. Annotate your diagram with the processes that occur within it. You can add more as you read the rest of this guide.

Synoptic links

The structure of amino acids is in Module 2. This would be a good opportunity to revise the structure and formation of peptide bonds in the synthesis of proteins. Liver cells make albumen, which is the major plasma protein that maintains the oncotic pressure (or water potential) of blood plasma.

This module covers metabolic pathways involving the biochemicals that you studied in Module 2. The ornithine cycle is the first of these. Later, you will learn about photosynthesis and respiration, which have linear, branched and cyclical pathways.

The kidneys

The kidneys are the main excretory organs, situated at the back and top of the abdomen just below the diaphragm. The kidneys filter blood and excrete waste products and substances in excess of requirements. They are the effector organs in water regulation.

At birth, each kidney contains a million unit structures known as nephrons. The number decreases with age. Each nephron filters blood to produce a liquid (filtrate) from the plasma, which contains useful substances as well as waste substances. Nephrons reabsorb useful substances into the blood and control the volume of water lost in the urine.

Figures 10, 11 and 12 show the structure of the kidney.

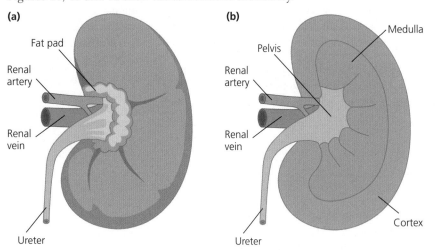

(a)
Fat pad
Renal artery
Renal vein
Ureter

(b)
Pelvis
Renal artery
Renal vein
Ureter
Medulla
Cortex

Figure 10 The gross structure of a kidney: (a) external view; (b) a vertical section

It is not possible to see complete nephrons in a kidney section because they do not lie in any one plane. Sections like those in Figures 11 and 12 show the parts of adjacent nephrons.

Exam tip

The nephron is the unit structure or functional unit of the kidney. We can explain the functions of the kidney by looking at what happens along the length of one nephron. This is because all nephrons are basically the same, differing only in the lengths of their loops.

(a)

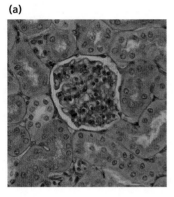

(b)

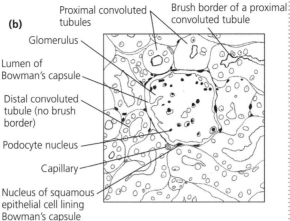

Proximal convoluted tubules

Glomerulus

Lumen of Bowman's capsule

Distal convoluted tubule (no brush border)

Podocyte nucleus

Capillary

Nucleus of squamous epithelial cell lining Bowman's capsule

Brush border of a proximal convoluted tubule

Figure 11 (a) A photomicrograph of part of the cortex of the kidney. (b) A drawing made from the photomicrograph

The cortex contains glomeruli, proximal convoluted tubules and distal convoluted tubules. Each glomerulus consists of a tightly arranged group of capillaries. These capillaries sit inside the cup-like structure called the Bowman's capsule (also known as the renal capsule).

(a)

(b)

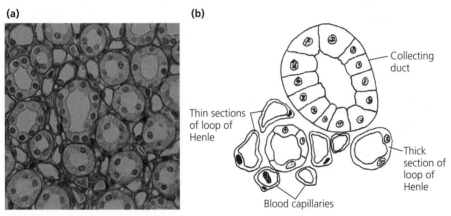

Collecting duct

Thin sections of loop of Henle

Thick section of loop of Henle

Blood capillaries

Figure 12 (a) A photomicrograph of part of the medulla of the kidney. (b) A drawing made from part of the photomicrograph

The inner part of the kidney is called the pelvis and is where urine collects. Between it and the cortex is the medulla, which contains loops and collecting ducts. These are visible in cross-section in Figure 12.

Figure 13 shows the structure of a single nephron and its associated blood vessels.

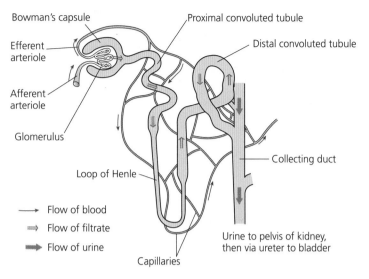

Figure 13 A kidney nephron. Blood from the renal artery flows into the afferent arteriole. Blood from the capillaries flows into the renal vein, to leave the kidney

Ultrafiltration

Blood enters the **glomerulus** from a branch of the renal artery at high pressure. Notice that the diameter of the efferent arteriole is narrower than that of the afferent arteriole. This builds up a 'head of pressure', which forces small molecules through the glomerular capillaries into the Bowman's capsule. This **pressure filtration** occurs in all capillaries, but is made more effective here by the arrangement of capillaries and podocytes (Figure 14). The endothelial cells lining the capillaries have numerous pores that allow substances to leave the blood and the podocytes have slit pores that reduce the resistance to the flow of filtrate.

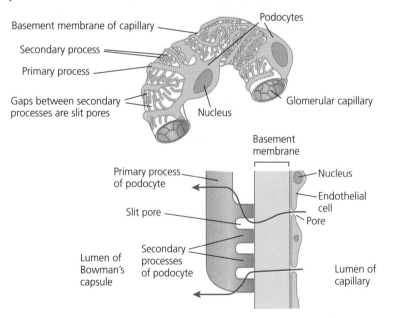

Figure 14 Ultrafiltration in the glomerulus

On the outside of the endothelial cells is a basement membrane made of glycoproteins. This acts as a molecular sieve, retaining cells and all molecules, such as proteins, with a relative molecular mass of more than 69 000, and allowing ions and small molecules, such as water, glucose and amino acids, to pass through. The capillaries in the glomerulus are supported by podocytes, which form an incomplete layer so they do not offer resistance to the flow of filtrate. The processes of the podocytes are arranged like loosely interlocked fingers. The spaces between the fingers represent the slit pores.

Selective reabsorption

Filtrate collects in the Bowman's capsule and passes into the proximal convoluted tubule (PCT). The cells that line the PCT are specialised for reabsorption of useful substances from the filtrate. Figure 15 shows the structure of these cells and how they reabsorb sodium ions and glucose (amino acids are reabsorbed by similar methods). Urea diffuses across the cells back into the bloodstream. The movement of solutes from the filtrate to the blood gives the blood a lower water potential than the filtrate. As a result, water passes by osmosis from the filtrate back into the blood. The reabsorption of urea and water is passive.

Absorption of glucose is an active process that requires a supply of ATP from respiration. ATP provides energy for sodium–potassium ion (Na^+/K^+) pump proteins on the lateral and basal membranes of the cells. The pumps create a low concentration of sodium ions inside the cytoplasm. This creates a concentration gradient for sodium ions from the filtrate into the cytoplasm, which is used to drive the uptake of other molecules, such as glucose and amino acids. Sodium ions can only diffuse through specialised protein channels or carriers.

The protein carrier on the luminal membrane facing the filtrate is a **symport** (co-transporter protein) that has binding sites for sodium ions and glucose. When both of these are filled, the carrier changes shape to deliver the sodium ion and glucose into the cytoplasm. This gives a high concentration of glucose inside the cell and glucose molecules diffuse out through carrier proteins in the basal and lateral membranes into the blood. The absorption of glucose and amino acids in this way is an example of **indirect active transport** — the molecules move into and out of the cell, but their movement is driven by the active transport of sodium ions and the presence of the co-transporter protein in the luminal membrane.

PCT cells are a good example of the relationship between structure and function:

- Tight junctions (acting like Velcro®) between cells ensure movement occurs through the cells, not between them.
- Microvilli give a large surface area for many symport carrier proteins.
- Mitochondria produce ATP for active transport by Na^+/K^+ pumps.
- Infoldings of the basal membrane give a large surface area for movement of substances into the blood.
- Rough endoplasmic reticulum makes proteins for Na^+/K^+ pumps, symport carrier proteins and glucose channel proteins.

Exam tip

Afferent and efferent are words often used in anatomy to describe structures, such as blood vessels, that 'go to' or 'come from' an organ or other structure. You need to refer to them when explaining ultrafiltration. Remember that **e**fferent = **e**xit.

Knowledge check 5

The volume of blood that passes through the glomeruli is 1200 cm³ min⁻¹. Of this, 125 cm³ forms filtrate. The plasma forms 55% of the blood by volume. Calculate the percentage of the plasma that forms filtrate every minute.

Exam tip

A tight junction holds two cell surface membranes close together so molecules cannot pass from the filtrate to the spaces between the cells that you can see in Figure 15.

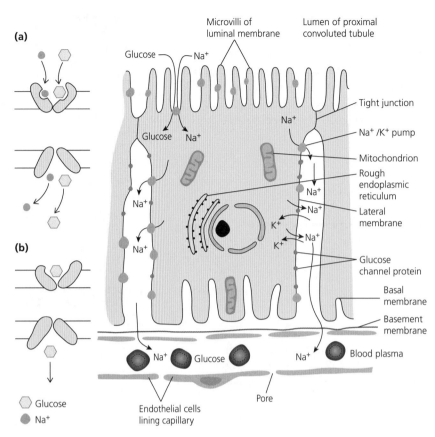

Figure 15 Selective reabsorption by a cell of the proximal convoluted tubule:
(a) movement of sodium ions and glucose through the luminal membrane;
(b) movement of glucose through the basal membrane

By the time the filtrate has reached the end of the PCT a large proportion of the solutes and much of the water has been reabsorbed. The remaining filtrate passes into the loop and then into the distal convoluted tubule (DCT).

Creating a water potential gradient

The main function of the rest of the nephron is to regulate the concentration of the blood and determine the concentration of the urine. The maximum concentration of urine in mammals is determined by the lengths of the loops and therefore by the depth of the medulla. This is achieved by making the tissue fluid in the medulla have a low water potential (with high concentrations of sodium ions and urea) so that water is reabsorbed by osmosis from urine in the collecting ducts.

Humans are able to concentrate urine by a factor of four compared with the concentration of blood plasma. This means that when dehydrating we can conserve water, rather than it going to waste in urine. Desert mammals with relatively longer loops than ours produce more concentrated urine. Some species can concentrate their urine up to a factor of 25 times the concentration of blood plasma.

As filtrate flows down the loops it becomes more concentrated. Water diffuses out into the surrounding tissue fluid as the walls are permeable to water, but have low

Knowledge check 6

State the changes to the filtrate as it flows through the PCT.

permeability to ions and urea. The filtrate flows up the loops after the hairpin bend. Here, the walls are impermeable to water, but solutes such as ions diffuse into the filtrate. Towards the top of the loops there are cells rich in mitochondria that pump ions out of the filtrate into the tissue fluid. It is this active transport that is mainly responsible for the very low water potential of the tissue fluid in the medulla.

Regulation of the water content of the blood

The water potential of the blood is a physiological factor that is kept within narrow limits. The hypothalamus is the control centre and contains osmoreceptors that detect changes in the water potential (oncotic pressure) of the blood. Cells do not function efficiently if their water content varies. If the water potential is too high, cells absorb water and swell, maybe even burst. If the water potential is too low, water diffuses out by osmosis and cells shrink.

Neurosecretory neurones from the hypothalamus pass into the posterior pituitary gland and terminate near blood capillaries (Figure 1d, p. 7). When they are stimulated, impulses travel down the axons to release molecules of a small peptide hormone called antidiuretic hormone (ADH) by exocytosis of vesicles. The target cells of ADH are the cells of the DCT and collecting ducts. Figure 16 shows what happens when ADH attaches to receptors on the surface membranes of these cells. Aquaporins are water channels that become inserted in the cell-surface membrane. When open, 3 billion molecules of water a second move through each aquaporin. (There are many different types of aquaporin. The type that responds to ADH is aquaporin 2.)

> **Exam tip**
>
> Make a large, simple diagram of the nephron with a long loop of Henle and a collecting duct. Draw in the blood vessels in the medulla as a loop. Notice that loops of Henle, blood vessels (known as the vasa recta) and collecting ducts are all parallel to one another.

> **Knowledge check 7**
>
> Aquaporins are selective for water and do not allow ions to pass through. Suggest how this happens.

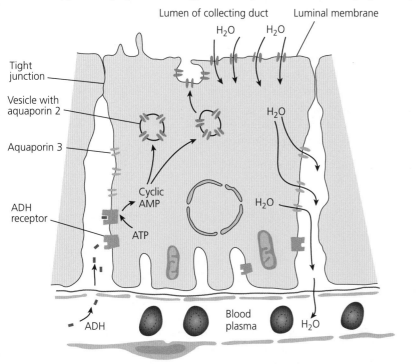

Figure 16 The changes that occur in a cell in a collecting duct in response to stimulation by ADH (aquaporin 1 is found in cell surface membranes of the PCT and descending limb of the loop of Henle)

The hypothalamus is also the body's thirst centre, so if the water potential decreases the thirst centre is activated and this stimulates the search for water and drinking.

When the water potential of the blood is higher than the set point, the osmoreceptors are not stimulated and ADH is not secreted. In the absence of ADH, aquaporins are taken back into the cytoplasm by endocytosis at the luminal surface. The membrane becomes impermeable to water and no water is reabsorbed by osmosis. Under these conditions, large volumes of dilute urine are produced. Osmoregulation is another example of negative feedback.

Synoptic links

There are many links to Module 2. Selective reabsorption is by active transport. There is also passive reabsorption of urea by diffusion and of water by osmosis. You may have to explain the control of water in the blood in terms of water potential. Aquaporins are protein channels that allow water to pass across membranes. Very little water can pass across the phospholipid bilayer, so aquaporins provide another route.

The vesicles with aquaporins are guided to the cell surface by the cytoskeleton. This also happens to vesicles in the synapse and in β-cells in islet tissue in the pancreas. In all three cases, the mechanism is the same — calcium ions stimulate the movement of the vesicles to the cell surface. This is an example of calcium ions acting as a second messenger.

Kidney failure

Kidneys can fail for a number of reasons, including blood loss. Kidney failure can prove fatal because urea, water, salts and toxins are retained and not excreted. With a decrease in blood pressure the **glomerular filtration rate** (GFR) decreases. The concentrations of ions, such as sodium, potassium, phosphate, hydrogencarbonate and chloride, are disturbed. The kidney has important roles in maintaining these concentrations and in kidney failure there is a loss of **electrolyte balance**.

Kidney failure can occur suddenly and last for a short time. In some cases, it is a long-term condition. Kidney failure can be treated by:

- dialysis — toxins, metabolic wastes and excess substances are removed by diffusion through the dialysis membrane (see below)
- kidney transplant

Renal dialysis

Dialysis means separating small and large molecules using a partially permeable membrane.

In haemodialysis, a partially permeable membrane in the machine separates the blood from a dialysis fluid (dialysate). Blood flows through tubes of dialysis membrane and the tubes are surrounded by dialysate. The dialysate contains substances required in the blood, such as glucose and sodium ions, in the correct concentrations. The dialysate does not contain any urea, so this diffuses from the blood down its concentration gradient into the dialysate, which then goes to waste.

Glomerular filtration rate The rate at which filtrate is formed by glomeruli in the kidneys. In humans the rate is constant, at about $125\,cm^3\,min^{-1}$.

Electrolyte balance The concentrations of physiologically important ions (e.g. Na^+, K^+ and Cl^-) in body fluids, such as blood and tissue fluid.

Dialysis fluid and blood flow in opposite directions so there is a concentration gradient along the whole length of the dialyser. Each time a unit volume of blood passes through the dialysis machine it loses some of its urea. After several hours, almost all the urea is removed. Haemodialysis involves treatment in hospital, in a clinic or at home at least three times a week.

The other form of dialysis is continuous ambulatory peritoneal dialysis (CAPD). Dialysate is placed in the abdominal cavity, and urea, other wastes and substances excess to requirements diffuse from the blood across the lining of the abdomen (the peritoneum) into the dialysate which is replaced regularly.

Diet has to be controlled carefully so that people do not produce too much urea or ingest too much salt.

A solution to these problems is to carry out a kidney transplant. Only one kidney is required, but the problem is finding a donor. The blood groups of donor and recipient must be compatible and there also needs to be a reasonably good match between the tissue types.

Urine tests

Urine tests using monoclonal antibodies (mAbs) are used to detect hormones, drugs of misuse and performance-enhancing drugs. These antibodies are manufactured by culturing modified B lymphocytes that secrete antibodies with antigen-binding sites complementary to the substance tested. They are also produced using genetically engineered cells.

The antibodies are either adsorbed onto the surface of wells in plastic plates or attached to sample test sticks. Enzymes that convert a colourless substrate into a coloured product are attached to the antibodies so that a colour change indicates a positive result.

Most performance-enhancing drugs are anabolic steroids, which stimulate the build-up of muscle mass because they stimulate protein synthesis. Athletes are tested regularly for these drugs. As these drugs circulate in the body they are filtered in the kidney and excreted. They are also broken down in the liver. This happens at different rates for different drugs. The length of time it takes for the concentration of a drug in the blood to decrease by a half is called its half-life. As many of these drugs have half-lives of about 16 hours it is important that urine samples are taken close to the event. Police and employers use similar kits to detect drugs of misuse, such as cannabis and cocaine.

These tests have been adapted for home testing kits for fertility and pregnancy tests. Antibodies specific to the hormone human chorionic gonadotrophin (hCG) are used in pregnancy tests. hCG is secreted by the embryo shortly after implantation, so its presence in urine indicates a positive result.

Exam tip

Hormones that are naturally produced in the body also have half-lives. The half-lives of insulin and glucagon are between 5 and 10 minutes and the half-life of adrenaline is 160–190 seconds.

Knowledge check 8

Insulin is detected in the urine. Suggest why this is so.

Synoptic links

Filtrate is similar to tissue fluid in that it has been filtered from the plasma through the walls of capillaries. In Module 3 you studied the differences between plasma, tissue fluid and lymph. You could draw up a table to compare the components of these three body fluids and urine.

In the UK, kidneys for transplant are usually matched exactly for blood group. You may already know that the ABO blood group is controlled by a gene with codominant alleles (if not, see Module 6). Kidneys may be rejected because of a mismatch in tissue types — it is rarely possible to match these exactly. The immune system of the recipient detects antigens on the surface of the kidney cells as being non-self, which stimulates an immune response resulting in destruction of the kidney cells by killer T cells (Module 4). People who have had a kidney transplant take immunosuppressive drugs to prevent the immune system from destroying the kidney.

Summary

- Excretion is the removal of metabolic waste and substances in excess of requirements.
- The liver receives blood from the hepatic artery and the hepatic portal vein; it is drained by the hepatic vein. Bile is stored in the gall bladder and passes to the duodenum in the bile duct. Liver cells (hepatocytes) are arranged into lobules in which each cell is in contact with blood.
- Excess amino acids are deaminated — the $-NH_2$ group forms ammonia, which is highly toxic. The reactions of the ornithine cycle combine ammonia and carbon dioxide to form the less toxic urea.
- The liver detoxifies drugs, such as alcohol, paracetamol and antibiotics.
- The kidneys are the main excretory organs. Blood is supplied in the renal artery and drained in the renal vein. Urine flows in the ureter to the bladder.
- The functional unit of the kidney is the nephron. Ultrafiltration occurs in the glomerulus; filtrate collects in the Bowman's capsule; selective reabsorption by active transport, diffusion and osmosis occur in the proximal and distal convoluted tubules. The loops of Henle and

collecting ducts maintain a high concentration of solutes in the tissue fluid in the medulla.
- The cortex is the outer region of the kidney; it surrounds the inner medulla and pelvis. The cortex has numerous glomeruli; the medulla has many loops, capillaries and collecting ducts that run in parallel.
- Osmoreceptors in the hypothalamus detect changes in water potential of plasma. When plasma becomes too concentrated, ADH is secreted by the posterior pituitary gland. Cells in the distal tubule and collecting ducts respond by inserting aquaporins in their cell membranes; water diffuses through these into tissue fluid in the medulla to conserve water.
- Treatment for people with kidney failure is regular dialysis or a kidney transplant.
- The hormone hCG is secreted from the very beginning of pregnancy and excreted in urine. Pregnancy tests use monoclonal antibodies to detect hCG. Performance-enhancing drugs and drugs of misuse are also detected by similar tests.

Neuronal communication

Key concepts you must understand

The organisation of the mammalian nervous system is described on pp. 47–49.

Nerve cells or **neurones** are specialised cells that transmit electrical impulses over long distances (Figure 1b, p. 7). One end of a nerve cell is specialised to receive information and the other end to send information. **Sensory neurones** (Figure 18a, p. 26) transmit impulses from receptors to the central nervous system (CNS). **Motor neurones** (Figure 18b, p. 26) transmit impulses from the CNS to muscles and glands.

Sensory receptors are **transducers** that convert one form of energy into another. Receptors are adapted to detect specific forms of energy and each type of receptor is adapted to detect a specific type of **stimulus**. Chemoreceptors in the nose and tongue have receptor molecules on their cell surface membranes that have shapes complementary to those of the molecules they detect. Rod and cone cells in the retina in the eye have light-sensitive pigments that absorb light energy. Pacinian corpuscles are pressure receptors in the skin (for details of how these function, see Question 6 on p. 78).

Key facts you must know

A **nerve** is a bundle of nerve cells or **neurones** surrounded by fibrous tissue. A section of a nerve is illustrated in Figure 17.

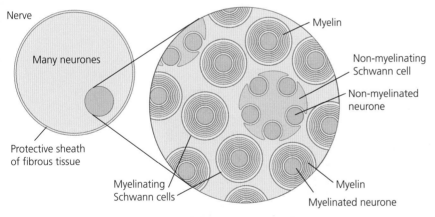

Figure 17 A cross-section of a nerve as seen using a light microscope

Neurones in nerves and in the spinal cord and brain are supported by **glial cells**. All the neurones in peripheral nerves are surrounded by special glial cells known as Schwann cells. Myelinated neurones are insulated by myelin made by Schwann cells and they transmit impulses very quickly. Non-myelinated neurones transmit impulses at much slower speeds (see Table 1 on p. 27).

Myelin is composed of concentric layers of cell membrane arranged like a Swiss roll. As they develop, Schwann cells wrap themselves around the neurone to give layers of thin sheets of cellular material (Figure 18c). Each layer of myelin consists

Neurone A nerve cell.

Transducer Anything that converts one form of energy into another.

Stimulus Any change in the environment.

Exam tip

The term *complementary* is often used. Remember what you learned in Module 2 about enzymes and substrates. There are many other examples of molecules 'fitting together'. Try listing them, as this will help your understanding of many aspects of biology.

Nerve A bundle of nerve cells surrounded by fibrous tissue.

Knowledge check 9

Explain why receptors are described as *transducers*.

Exam tip

You may not have heard of glial cells before, although you should have heard of Schwann cells. Glial cells have important roles other than providing protection and insulation. They remove neurotransmitters from synapses and maintain potassium ion concentrations in tissue fluid around neurones.

of two membranes with a very thin layer of cytoplasm sandwiched between them. The membranes are composed mostly of phospholipids; there are very few proteins. The inner layer of myelin is so close to the neurone that tissue fluid and the ions it contains do not reach its cell surface membrane. Tissue fluid only makes contact with the cell surface membrane at the **nodes of Ranvier**, which are gaps between the Schwann cells.

> ### Knowledge check 10
>
> The speed of conduction of neurones varies between $1\,m\,s^{-1}$ and $100\,m\,s^{-1}$. Explain briefly why there is such a wide range in conduction speeds.

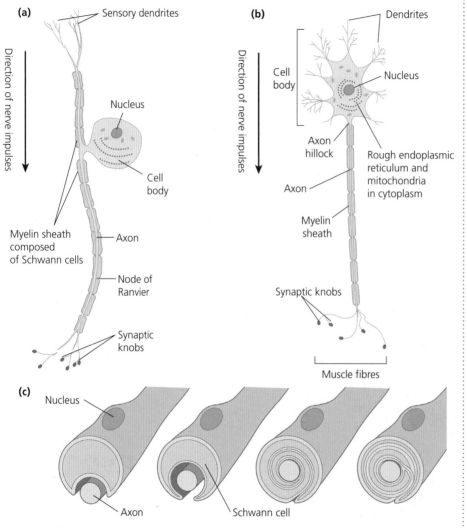

Figure 18 (a) A sensory neurone. (b) A motor neurone. (c) A Schwann cell making myelin by wrapping around a neurone

The cell bodies of sensory neurones are in swellings (ganglia) situated just outside the spinal cord. The cell bodies of many motor neurones are inside the spinal cord. Some motor neurones in the involuntary nervous system have their cell bodies in ganglia outside the spinal cord, for example in the solar plexus.

Table 1 Myelinated and non-myelinated neurones

Feature	Myelinated neurones	Non-myelinated neurones
Schwann cells surrounding neurones	Produce myelin, comprising concentric layers of membrane rich in lipid that act as insulation, preventing ions from reaching the neurone cell surface membrane	No myelin produced
Action potentials	Occur only at nodes of Ranvier	Occur along the whole length of the neurone
Conduction of impulse	Saltatory ('jumping')	Contiguous — wave-like, along the whole neurone
Typical speed of impulse transmission/m s^{-1}	100–120	10–20
Distribution in animal groups	Vertebrates (a few groups of invertebrates have myelinated neurones)	Invertebrates (e.g. squid and earthworm); vertebrates
Roles in mammals	Impulse transmission in the voluntary nervous system, for example in controlling contraction of skeletal muscles in walking; rapid transmission in the CNS, for example up and down the spinal cord	Impulse transmission in the involuntary system, for example controlling the heart rate

Neurones do not bring about behaviour on their own — they work in circuits. The simplest piece of behaviour we can show is a reflex action, which is coordinated by two or three neurones working in series. Figure 19 shows the arrangement of three neurones that control a reflex such as those described on pp. 51–52. Relay neurones are found only within the grey matter of the spinal cord and the brain. They allow connections within the nervous system that may influence the reflex action, for example inhibit it (p. 36).

Knowledge check 11

Distinguish between the roles of sensory, motor and relay neurones.

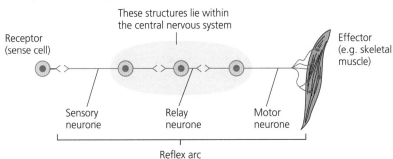

Figure 19 Three neurones that control a reflex action

Transmission of impulses

A **nerve impulse** is the flow of current along a neurone. This is similar to current flow in a wire, except that it does not involve the flow of electrons. Current decays because of the resistance provided by cytoplasm and by the leakage of positive ions out through the membrane into tissue fluid.

In very short neurones, such as those in the brain and in the retina of the eye, flow of current is sufficient to carry information along the length of the neurone. If the

Exam tip

It is well worth finding some good animations of the conduction of nerve impulses so you can appreciate the importance of action potentials and how they are propagated along neurones. Take notes as you watch them.

Nerve impulse A flow of current along the length of a neurone.

distance is more than 1–2 mm the current has to be 'boosted' at intervals along the neurone. This 'boosting' effect is an **action potential** and is how an impulse is propagated along a neurone. In non-myelinated neurones boosting occurs all the way along the neurone to give **contiguous** flow. For an impulse to travel the length of a neurone, action potentials must occur at intervals. The word propagation means 'produce more' and is used to mean the production of action potentials that are repeated all along the neurone.

In myelinated neurones, the 'booster' effect occurs only at the nodes of Ranvier where the axon membrane is exposed to the tissue fluid. Impulse transmission in myelinated neurones is **saltatory** ('jumping') and is faster than contiguous transmission in non-myelinated neurones.

Exam tip

The 'boosting' is achieved by using the energy stored in gradients for sodium and potassium ions. The inward flow of sodium ions depolarises the neurone and gives the 'push' needed for current flow; the outward flow of potassium ions restores the **resting potential**. Remember these principles as you read on.

Figure 20 shows the important components of the membrane of an axon. Leak channel proteins are open all the time, but selective for sodium and potassium ions. There are many more of these channels for potassium. Voltage-gated ion channels are shut when the axon is not conducting impulses. When stimulated, the activation gates open to let ions flow through the membrane. The gates on sodium channels open faster than those on potassium channels. Voltage-gated sodium ion channel proteins have an additional inactivation gate that closes and stays shut for a while before opening again. Pump proteins use ATP to pump sodium ions out and potassium ions into the axon.

Action potential The increase and decrease in potential difference across a neuronal membrane when it is stimulated.

Contiguous Close together. Action potentials in non-myelinated neurones occur all along the axon. In myelinated neurones action potentials occur only at nodes of Ranvier.

Saltatory Refers to action potentials that occur only at nodes in myelinated neurones, and explains why these neurones have high speeds of conduction.

Resting potential The potential difference across a neuronal membrane when it is not stimulated.

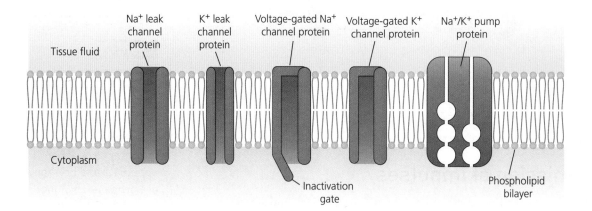

Figure 20 The neuronal membrane contains leak channel proteins and voltage-gated channel proteins for facilitated diffusion, and sodium–potassium pump proteins for active transport. Both types of voltage-gated channel protein have activation gates, but only the sodium channel protein has an inactivation gate

Respiration provides ATP to pump ions and maintain concentration gradients. This is like a constantly recharging battery.

The effect of these pumps and channels can be seen when an impulse passes along a neurone. The electrical activity in neurones is detected by placing electrodes at intervals across a neuronal membrane (Figure 21). The electrodes detect changes in potential difference across the membrane as the impulse passes. The impulse is detected as a wave of depolarisation in which the potential difference across the membrane is reversed. Immediately afterwards, the neurone is repolarised, returning to its resting value.

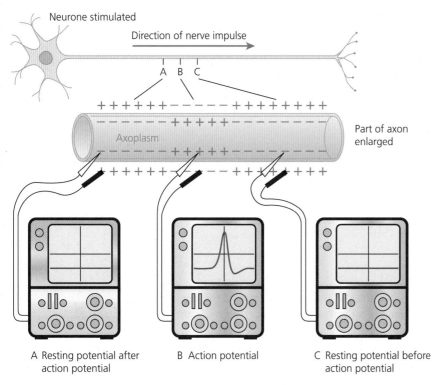

Figure 21 Changes in potential difference across neuronal membranes are detected using microelectrodes and displayed on screen

Resting potential

When not sending an impulse, a neurone is at a **resting potential**. The resting potential is caused by the unequal distribution of ions across the membrane. This happens in all cells, including plant cells, and is a property of cell surface membranes. Cytoplasm contains organic anions (e.g. negatively charged proteins) that cannot diffuse through the membrane because they are charged and are too big.

The pump proteins shown in Figure 20 pump three sodium ions out for every two potassium ions pumped in. The membrane is relatively impermeable to sodium ions because they are charged and cannot pass through the hydrophobic region of the phospholipid bilayer, and because there are few leak channel proteins specific for these ions. Voltage-gated channel proteins for sodium ions and potassium ions are shut when neurones are at rest.

The potential difference across the membrane is between −60 mV and −70 mV. This means that the *inside* of the membrane is negatively charged with respect to the *outside*.

The actual resting potential depends on the concentration of potassium ions in the tissue fluid around the neurone. There are more leak channel proteins for potassium ions than for sodium ions and as they are open all the time they allow potassium ions to flow out down their concentration gradient. However, the internal negative charge attracts many potassium ions, so they tend to remain inside the neurone, giving a high intracellular K^+ concentration.

At this stage, the neurone can be compared to a battery before current starts to flow in the circuit. The neurone has a store of energy in the form of concentration gradients for ions. In addition, the inside of the neurone has a negative charge so there is an electrical gradient attracting positive sodium ions into the neurone. The sum of these two gradients is called the **electrochemical gradient**. There is also a concentration gradient for potassium ions as there is a higher concentration of these inside the neurone. These gradients are put to work during the passage of an impulse.

> **Exam tip**
>
> When you write about a nerve impulse or describe the events that occur during an action potential always start by referring to the resting potential. It may not be appropriate to explain how it is achieved, but you should refer to its value of about −70 mV.

Action potential

There are special channels in the axon hillock of a motor neurone that open in response to the stimuli received from other neurones. If the stimulus is great enough, the axon hillock is depolarised sufficiently so that an **action potential** occurs. This generates current flow along the neurone and gives a higher concentration of positive ions in the region immediately adjacent to the axon hillock. The membrane is depolarised as the potential difference is reduced from −70 mV to −50 mV.

Voltage-gated sodium ion channel proteins are sensitive to this depolarisation. The arrival of positively charged ions is enough to trigger the activation gates of some of these channel proteins to open, allowing sodium ions to flow down their electrochemical gradient across the membrane *into* the neurone. As this happens, the potential difference becomes less negative. This triggers more of these sodium channels to open. The potential difference continues to change, triggering yet more to open. This is a **positive feedback** in that a disturbance stimulates more and more channels to open, causing an explosive change in the potential difference. This only happens for a very short time and not many ions flow in. Voltage-gated sodium ion channel proteins remain open for a short while and then shut firmly as their inactivation gates close. The potential difference reaches +30 mV to +40 mV (Figure 22a).

To return the potential difference to −70 mV, voltage-gated potassium ion channel proteins respond to the change in potential difference and open, allowing potassium ions to flow out of the neurone down their concentration gradient. This restores the resting potential very quickly.

> **Exam tip**
>
> You may see different values given for the resting potential. It varies between neurones, but is usually between −60 mV and −70 mV.

> **Electrochemical gradient** A gradient for charged particles (e.g. sodium ions) across a membrane composed of both the concentration gradient and the gradient of electrical charge.

> **Exam tip**
>
> You may not have come across the term axon hillock before. It is rarely, if ever, mentioned in school textbooks. See Figure 18b to see where it is — at the junction of the axon with the cell body.

> **Knowledge check 12**
>
> State the maximum change in the potential difference across a neurone membrane during an action potential.

Figure 22 shows that the high conductance for sodium ions coincides with the rising phase of the action potential. This is when the voltage-gated sodium ion channel proteins are open and sodium ions diffuse into the axon down an electrochemical gradient. During the millisecond that the channels are open some 7000 sodium ions enter a neurone. The falling phase coincides with the inactivation of these channel proteins when they shut. It also coincides with the high conductance for potassium ions when the voltage-gated potassium ion channel proteins open, allowing potassium ions to diffuse out and the membrane to repolarise. There is an 'undershoot' to -90 mV because these channel proteins are slower to close than the sodium channel proteins; more potassium ions diffuse out than necessary to return the potential difference to -70 mV.

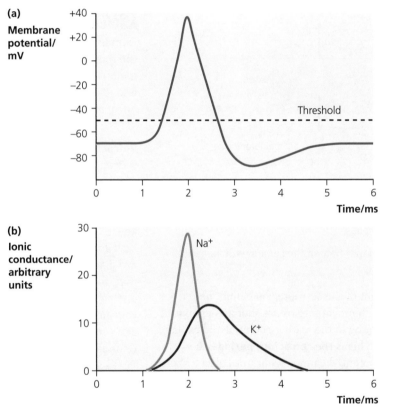

Figure 22 The change in (a) potential difference and (b) conductance of sodium and potassium ions during the passage of an action potential

Propagating the action potential

At the height of the action potential there is a greater concentration of positively charged ions inside the neurone. These boost the current flow to the next part of the neurone. Current flow is illustrated in Figure 23. The impulse is shown travelling from right to left. The active region is the part of the axon at the height of the action potential with a membrane potential of $+40$ mV. The inflow of sodium ions increases the concentration of positively charged ions (mainly K^+). These repel each other and are attracted by large negatively charged anions, so they spread outwards to left and right inside the axon.

> **Exam tip**
>
> Voltage-gated *sodium* ion channel proteins are more sensitive to depolarisation than voltage-gated *potassium* ion channel proteins. The latter open more slowly which explains the conductance of the two ions that you can see in Figure 22b.

> **Exam tip**
>
> Figure 22 shows the changes that occur at one place on an axon. The ions flow transversely — they flow across the membrane, not along it. Electrodes at this place will detect many such action potentials per unit time; the frequency is determined by the axon hillock.

> **Exam tip**
>
> Copy Figure 22 onto the centre of a large piece of paper. As you read the next few pages, annotate the figure with bullet points to make a good revision aid.

On the outside of the axon there is a slight negative charge as sodium ions have moved into the axon. This attracts positively charged ions (mainly Na⁺) from both sides, which flow in the tissue fluid. This flow of current (as positively charged ions on both sides of the membrane) depolarises the next patch of membrane to the left. This depolarisation is the beginning of the rising phase of the action potential and will always reach the threshold value to stimulate the opening of many voltage-gated sodium channels.

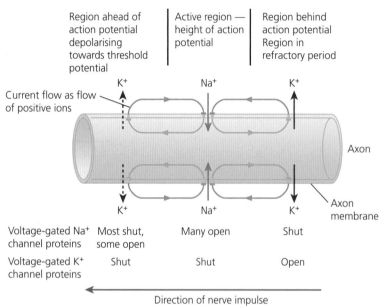

Figure 23 Current flow in a neurone in front of and behind the region with an action potential

Meanwhile, the part of the neurone to the right of the active region cannot be stimulated because the voltage-gated sodium channel proteins are shut and cannot be opened. This happens while this area is returning to the resting potential and explains why impulses travel one way along neurones. This is the **refractory period**, when depolarisation cannot occur. Another action potential cannot occur until after a short gap, so making impulses discrete events.

Remember that action potentials boost current flow, so an impulse can travel along a neurone without decaying. Action potentials do not vary in size — they all show the same amplitude (or 'spike'). However, the **frequency** at which they are sent along neurones varies. If the stimulus is large, there is a high frequency of impulses. A lower stimulus leads to a lower frequency of impulses. So at **A** on Figure 21 the screens would show a large number of impulses per second with a large stimulus, but a small number per second with a low stimulus.

A second stimulus applied to a neurone less than 1 ms after the first will not trigger another impulse. The membrane is depolarised and the neurone is in its refractory period. Not until the −70 mV polarity is re-established will the neurone be ready to fire again. Repolarisation is established by the facilitated diffusion of potassium ions *out of the neurone*. In some human neurones, the refractory period lasts only 1–2 ms. This means that these neurones can transmit up to 500 to 1000 impulses per second.

Exam tip

Why always? Remember that once an action potential starts at the axon hillock it is repeated all the way down the neurone — an **all-or-nothing response** (pp. 78–79).

Exam tip

Remember that Na⁺ is the main extracellular cation (in tissue fluid) and K⁺ is the main intracellular cation.

Refractory period
Refractory means stubborn or unresponsive. The refractory period is the key to understanding how action potentials are discrete events; they do not merge into one another. Drum your finger on a table at a regular frequency. Change the frequency. There's your model of information encoding by neurones.

Knowledge check 13

Explain why a nerve impulse travels in one direction along a neurone.

The events that occur in a neurone membrane when an impulse is transmitted are shown in Figure 24. This also shows the movement of ions through a sodium–potassium ion pump protein that occurs in neurones all the time.

Knowledge check 14

Use Figures 22 and 24 to describe the changes that occur to the voltage-gated ion channels in an axon membrane during the passage of an action potential. You may find it easiest to use a table for your answer.

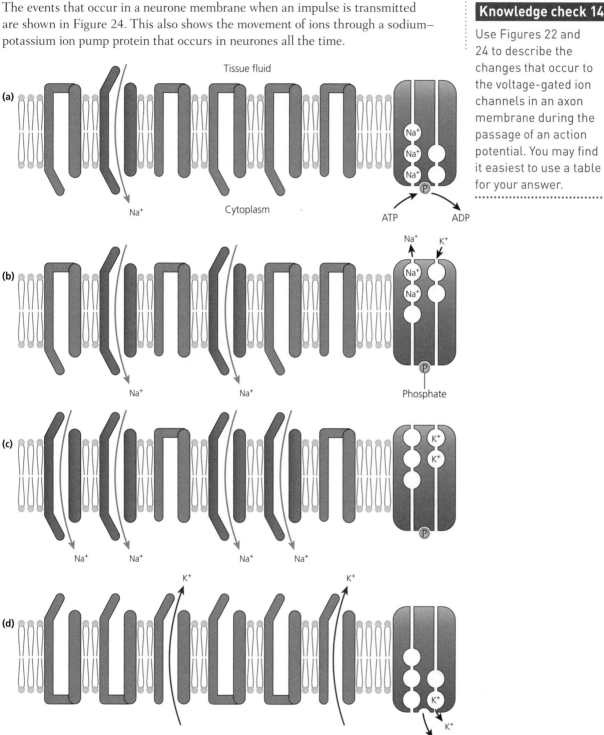

Figure 24 Ion transport at different stages of an action potential: (a) during depolarisation towards threshold; (b) as voltage-gated sodium ion channels begin to open in the rising phase of an action potential; (c) at the height of an action potential; (d) during repolarisation

Sometimes the stimulus received by a neurone is too low and no impulse is sent. This is because the depolarisation of the membrane of the cell body is not great enough to trigger an action potential at the axon hillock. For impulses to be sent, the depolarisation must be above a **threshold** value, which is usually 10–15 mV above the resting potential. If the depolarisation is above threshold, then an impulse is sent. Neurones either transmit impulses or they don't — this is known as the **all-or-nothing response**. They do *not* transmit action potentials of different amplitudes in response to different intensities of stimulus.

The voltage-gated ion channels in myelinated neurones are concentrated at the nodes and are not in the membrane covered by myelin. This saves energy, since fewer ion channels need to be made and pump proteins are only active at the nodes. Speeds are much faster as far fewer action potentials are needed in comparison with an unmyelinated neurone where they occur contiguously. If a node is inactive for some reason, then the local currents shown in Figure 22 will stimulate the next node. Local currents decay, so this will not work if two or more nodes are inactive.

Information is encoded in the nervous system in the following ways:

■ Axon hillocks send impulses if the stimulus is above the threshold. Different neurones have different threshold values.
■ Neurones follow distinct pathways into and out of the brain and spinal cord. Sensory neurones terminate on specific cells in the CNS that interpret impulses as coming from specific receptors. Motor neurones for specific effectors originate in specific parts of the CNS.
■ The number of receptors that respond and send impulses indicates the extent of the area stimulated.
■ The frequency of impulses indicates the strength of the stimulus.

Synoptic links

The sodium–potassium ion pump is an example of active transport. This is an **antiport** transport protein that pumps ions in opposite directions across the membrane. It is also an example of a **direct active transport** system as the pump protein uses ATP. Compare this with the indirect active transport using the symport transport protein in PCT cells (p. 19). The movement of sodium and potassium ions through their voltage-gated channel proteins is an example of facilitated diffusion.

Synapses

Neurones make contact with their target cells at synapses. The term synapse comes from the Greek for 'fastening together' and it applies to the gap between neurone and target cell and the areas either side of the gap. In some synapses there is no gap and impulses pass from one neurone to another electrically. We have these electrical synapses in our brains. In the synapses that we are considering, impulses are transmitted chemically using neurotransmitter substances. One of the most common neurotransmitters is **acetylcholine** (ACh). Figure 25 shows the structure of a typical synapse, which could be in the brain or spinal cord or in swellings called ganglia along some nerves. Synapses with acetylcholine as the neurotransmitter are called **cholinergic synapses**.

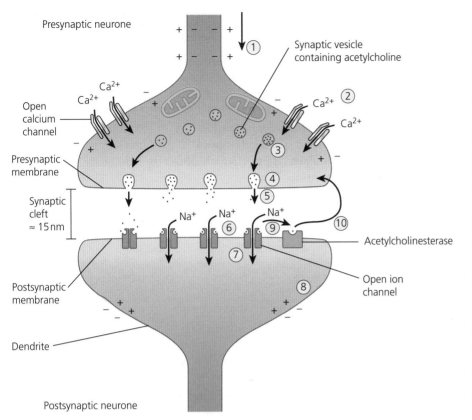

Figure 25 The events that occur during the transmission of an impulse across a cholinergic synapse

1 An action potential arrives at the end of the presynaptic neurone.

2 Depolarisation stimulates voltage-gated calcium ion channels to open.

3 Calcium ions diffuse through the voltage-gated calcium ion channel proteins.

4 Calcium ions stimulate movement of synaptic vesicles towards the presynaptic membrane.

5 Vesicles fuse with the presynaptic membrane and release acetylcholine molecules by exocytosis into the synaptic cleft. Each vesicle contains about 10 000 molecules of acetylcholine.

6 Acetylcholine molecules diffuse across the synaptic cleft and attach to chemical-gated ion channels (this diffusion takes under 1 ms to occur).

7 Chemical-gated ion channel proteins open and sodium ions diffuse across the postsynaptic membrane. These are also known as ligand-gated channels.

8 Influx of sodium ions depolarises the postsynaptic membrane. The depolarisation of the postsynaptic membrane is known as the excitatory postsynaptic potential (EPSP).

9 Acetylcholine molecules leave the ion channels and enter active sites of the enzyme acetylcholinesterase, which hydrolyses each acetylcholine molecule, separating the acetyl (ethanoyl) group from choline. This enzyme removes molecules of acetylcholine very quickly so stimulation is a brief event.

10 Choline molecules enter the presynaptic membrane through transporter proteins and are resynthesised into acetylcholine using ATP from mitochondria.

The arrival of one impulse stimulates the release of acetylcholine by exocytosis. Since many impulses may arrive at a synaptic bulb at any one time they have an additive effect. Whereas one impulse does not stimulate the next neurone as the EPSP is not above threshold, the arrival of several impulses over a short period of time will depolarise the postsynaptic neurone sufficiently. This is known as **summation**. If it occurs along one neurone within a certain period of time it is called **temporal summation**. If impulses from different neurones arrive along different presynaptic neurones at the same time it is known as **spatial summation**.

The release of some neurotransmitters leads to hyperpolarisation — the membrane potential becomes more negative, making it more difficult to depolarise the membrane. This hyperpolarisation is an inhibitory postsynaptic potential (IPSP).

Roles of synapses

Synapses have a number of roles in communication:

- They permit impulses to travel from one neurone to another in one direction only. Neurotransmitters are only released on the presynaptic side and their receptors occur only on the postsynaptic side.
- Excitatory synapses cause an EPSP. If this depolarisation is above threshold, then an impulse is sent in the postsynaptic neurone.
- Inhibitory synapses cause an IPSP. The hyperpolarisation makes it more difficult for excitatory neurones to raise the postsynaptic potential to reach threshold. A common neurotransmitter at inhibitory synapses is gamma butyric acid (GABA).
- They filter information. If not above threshold, then the information is not sent any further. It is often *changes* in the intensity of stimuli that are detected.
- Synaptic knobs run out of neurotransmitter and so develop fatigue. This stops the presynaptic neurone from stimulating the postsynaptic neurone and protects against overstimulation.
- They integrate information from different neurones. A single relay or motor neurone may have many thousands of neurones that form synapses over its dendrites and cell body. In this way information from many places in the body can be integrated into a single response.
- Memory is a function of the interconnections between neurones in the brain.

Synoptic links

Neurotransmitters occur in vesicles that fuse with the presynaptic membrane and are released by exocytosis. They diffuse across the synaptic gap and interact with protein receptors on the postsynaptic membrane. This is an example of cell signalling. The receptor sites on the chemically gated channel proteins are specific for the type of neurotransmitter, as is the active site of the enzyme that hydrolyses acetylcholine. These are yet more examples of the specificity of proteins. In the autoimmune disease myasthenia gravis antibodies are produced against acetylcholinesterase.

Knowledge check 15

Outline what happens following the arrival of an action potential at the presynaptic membrane and the depolarisation of the postsynaptic membrane.

Exam tip

Many drugs have their effects by interacting with membrane proteins at synapses. Agonists mimic the effects of the neurotransmitter; antagonists work by blocking the chemical-gated channels. There are also drugs that inhibit acetylcholinesterase. Work out the effects of these three types of drug.

Summary

- The nervous system comprises neurones and glial (supporting) cells. Sensory neurones transmit impulses from receptors to the central nervous system (brain and spinal cord). Motor neurones transmit impulses from the CNS to effectors such as muscles and glands. Relay neurones transmit impulses between sensory and motor neurones.
- Sensory receptors are transducers; they convert different forms of energy (e.g. light, kinetic, chemical) into nerve impulses.
- The potential difference across cell membranes is a result of unequal distribution of ions, with more anions within the cell and more cations outside. In neurones, this is the resting potential. It is maintained by the sodium–potassium pump, using energy from ATP.
- Neurones transmit nerve impulses over long distances by using reversals of the potential difference to 'boost' current flow. The opening of voltage-gated Na^+ channel proteins allows inflow of sodium ions that depolarises the membrane from -70 mV to $+40$ mV. The opening of voltage-gated K^+ channel proteins allows outflow of potassium ions to restore the resting potential.
- Sensory and motor neurones are protected by Schwann cells, some of which make myelin, which acts as an insulator. Action potentials occur all along unmyelinated neurones. In myelinated neurones they occur at the nodes of Ranvier between Schwann cells; this gives a much faster saltatory conduction.
- Action potentials are propagated by current flow that spreads along a neurone to form local circuits that depolarise the next area of membrane to the threshold potential (e.g. -50 mV), which leads to the rising phase of an action potential.
- Action potentials are generated only if the threshold for the neurone is exceeded (all-or-nothing response). The strength of a stimulus is encoded as the frequency of impulses transmitted by a neurone.
- Neurones communicate with each other at synapses. Chemical signalling occurs at synapses by release of neurotransmitter chemicals, such as acetylcholine (ACh).
- Synapses permit one-way transmission of impulses in the nervous system and interconnections between many neurones. Weak impulses are filtered out if they are below the threshold of the postsynaptic neurone; inhibitory neurones stimulate hyperpolarisation in the postsynaptic neurone, making it more difficult to reach threshold.

Hormonal communication

Key concepts you must understand

Hormones are signalling molecules that travel long distances in the blood. They are secreted by endocrine (ductless) glands. The adrenal gland (Figure 26) is an example.

An adrenal gland is composed of an outer adrenal cortex that surrounds an inner adrenal medulla. The hormones from the adrenal cortex are steroids. Adrenaline and noradrenaline secreted by the adrenal medulla are catecholamines, which are made from amino acids.

Adrenaline, like insulin and glucagon (p. 41), cannot cross cell membranes, whereas lipid-soluble steroid hormones, such as testosterone, 17β-oestradiol and progesterone, can. These steroid hormones interact with receptors inside the cytoplasm or nucleus.

Hormone A chemical signal secreted by an endocrine gland into the blood so it is transported to its target cell.

Endocrine gland A ductless gland that secretes hormones into the blood.

Target cell A cell with receptors for a specific signalling molecule, for example a hormone, so it detects and responds to that signal.

Key facts you must know

Adrenaline

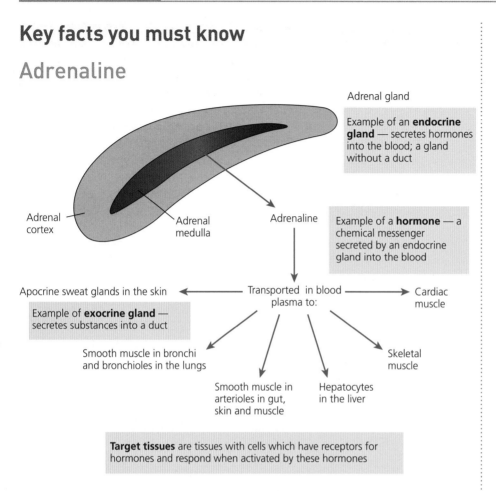

Figure 26 Adrenaline is secreted by the adrenal medulla and influences many target tissues throughout the body

Table 2 The roles of the adrenal cortex and the adrenal medulla

Region of the adrenal gland	Hormones secreted	Target organs	Functions of hormones
Adrenal cortex	Mineralocorticoid: aldosterone	Kidney, gut	Stimulates increase in absorption of sodium ions, leading to an increase in blood pressure
	Glucocorticoids: cortisol and corticosterone	Liver	Stimulate increase in blood glucose concentration by gluconeogenesis
Adrenal medulla	Catecholamine: adrenaline (epinephrine*)	Liver	For example: stimulates breakdown of glycogen to increase blood glucose concentration
	Catecholamine: noradrenaline (norepinephrine*)	Heart	Increases heart rate
*Adrenaline and noradrenaline are known by international agreement as epinephrine and norepinephrine.			

Adrenaline has numerous effects on the body. It is usually released in times of danger and stress and is often called the 'fight or flight' hormone. It prepares the body for meeting these situations by:

- increasing heart rate and stroke volume (volume of blood per beat)
- stimulating enzymes in the liver to convert glycogen to glucose, so releasing glucose into the blood
- decreasing blood flow to the gut and skin by stimulating vasoconstriction
- increasing blood flow to muscles and the brain by stimulating vasodilation
- increasing blood pressure because vasoconstriction increases resistance to flow
- increasing the width of bronchioles by causing smooth muscles to relax, so increasing air flow to the alveoli
- stimulating contraction of the radial muscles in the iris of the eyes to dilate the pupils
- increasing secretion from apocrine sweat glands that open into hair follicles (these glands are partly responsible for body odour)

First and second messengers

Cells in target tissues for adrenaline have specific **adrenergic receptors** on their cell surface membranes. Adrenaline molecules fit into these receptors and this causes a chain of events to occur inside the cell, as shown in Figure 27. The G protein acts as a transducer to transfer the message to the cytoplasm by activating the enzyme adenylyl cyclase, which converts ATP into **cyclic AMP**.

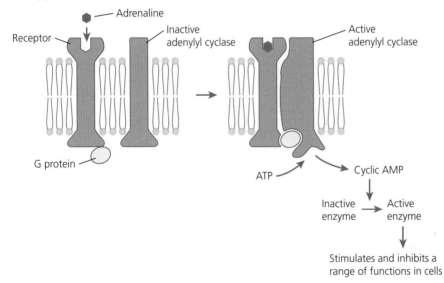

Figure 27 The role of the second messenger, cyclic AMP, inside cells stimulated by adrenaline

Adrenaline is a hormone, a **first** (or **primary**) **messenger**. Cyclic AMP is a **second** (or **secondary**) **messenger** that acts on enzyme systems within the cell. It interacts with a kinase enzyme that activates other enzymes in a cascade to amplify the original message. Many enzymes are activated in the process; some are inactivated. In liver cells, the enzyme glycogen phosphorylase is activated to remove glucose from glycogen. Glucose diffuses out of the cell into the blood plasma to increase the concentration as it is needed by muscles to contract during exercise or during 'fight or flight'.

> **Exam tip**
>
> Adrenaline works together with the nervous system to increase the readiness of the body for 'flight or fight'. Remember to write about both in answers to questions on responses to dangerous situations.

> **Exam tip**
>
> Cells are primed, ready to respond to the first messenger. Different types of cells are primed in different ways, which is why adrenaline stimulates very different responses in its target cells.

> **Exam tip**
>
> There are other second messengers. Calcium ions play an important role in plants and animals as second messengers. The concentration in the cytoplasm is very low — almost nothing. As soon as calcium ions enter, as at a pre-synaptic membrane, they stimulate events to occur, such as movement of vesicles.

The pancreas

The pancreas is in the abdomen just below the stomach and to the right of the liver when viewed from the front. It is both an endocrine organ and an exocrine organ:

■ Endocrine function — secretion of the hormones insulin and glucagon into the blood by cells in the islets of Langerhans; α-cells secrete glucagon; β-cells secrete insulin.

■ Exocrine functions — secretion of amylase, proteases, lipase and nucleases from acinar cells into the pancreatic duct. This duct empties into the duodenum, which is the first part of the small intestine (Figure 4, p. 12). The alkali sodium hydrogencarbonate is also secreted to neutralise the acid contents of the stomach as they enter the duodenum.

Knowledge check 16

Explain the difference between exocrine secretion and endocrine secretion.

(a)

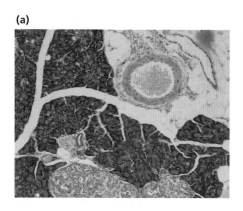

(b)

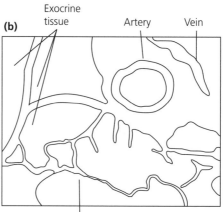

Figure 28 (a) Low-power photomicrograph of pancreatic tissue. (b) A plan drawing made from the photomicrograph

Exam tip

In your practical work you may have to make plan drawings (Figure 28) and high-power drawings (Figure 29) from slides of the pancreas. Download images like those here and practise making such drawings.

(a)

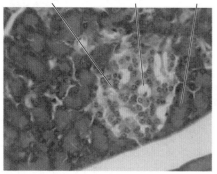

(b)

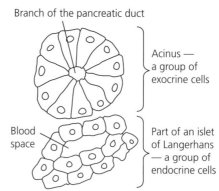

Figure 29 (a) High-power photomicrograph of the endocrine and exocrine areas of the pancreas. (b) A drawing showing cells from the two areas

Controlling the concentration of blood glucose

The glucose concentration in the blood fluctuates. As blood flows through capillaries, glucose is forced out into the tissue fluid. It is important that the concentration in the tissue fluid around cells is maintained so that cells have a constant supply.

If the concentration of glucose rises too high it cannot all be reabsorbed by the kidneys and is excreted in the urine — the concentration is said to exceed the renal threshold. This suggests that glucose is not being stored either as glycogen or as fat and energy reserves are not being topped up.

If the glucose concentration falls too low, then there is not enough to supply brain cells, which cannot respire anything other than glucose, and a person may enter a coma.

The glucose concentration in the blood is kept within the range $80-120 \, \text{mg} \, 100 \, \text{cm}^{-3}$ blood and is normally about $90 \, \text{mg} \, 100 \, \text{cm}^{-3}$, which we can take as the set point.

As a meal is absorbed into the bloodstream the concentration of glucose can increase by 50%. The following events occur when the blood glucose concentration rises above the set point:

1 The increasing concentration of glucose acts as a stimulus that is detected by β-cells in the islets of Langerhans, which release insulin in response.
2 Insulin circulates in the bloodstream and binds to insulin receptors on target cells. The insulin receptor is a transmembrane protein in the cell surface membrane of these target cells in muscle, liver and adipose (fat) tissue.
3 Insulin stimulates muscle and adipose cells to take up glucose and convert it into glycogen and fat. Insulin stimulates more channel proteins for glucose (known as GLUT4) to move into the plasma membranes of muscle and fat cells.
4 Insulin has a number of effects on liver cells including:
   – increasing the use of glucose, for example in respiration
   – stimulating the conversion of glucose into glycogen — this process is called **glycogenesis** (literally, 'making glycogen')
   – inhibiting the breakdown of glycogen to glucose
   – inhibiting the conversion of fats and proteins into glucose

   Using up glucose inside liver cells increases the uptake of glucose from the blood through channel proteins (known as GLUT1) that are always in the membrane and the numbers of which are not increased in response to insulin.
5 Glucose is now being stored or 'put away' for later. It is converted to glycogen for short-term storage and converted to fat for long-term storage. The concentration of glucose in the blood decreases.

After a meal has been absorbed completely and also during exercise the blood glucose concentration may decrease below the set point. The following events then occur:

1 β-cells stop releasing insulin. This means that cells take up less glucose.
2 α-cells in the islets of Langerhans respond to decreasing concentrations of glucose by releasing **glucagon**. (Glucagon has a paracrine effect on β-cells, stimulating them to make insulin so they can release it when the glucose concentration increases.)

Exam tip

Look back to the section on temperature control for the terms used when writing about homeostatic mechanisms. Set point is also called the 'norm', from the idea that it is the normal level.

Exam tip

Insulin has a paracrine effect on α-cells, stopping them from releasing glucagon so that glucose is not released into the blood by liver cells. (Paracrine secretion is illustrated in Figure 1a, p. 7.)

Exam tip

Insulin and glucagon are hormones; they are *not* enzymes. Avoid writing things like 'insulin converts glucose to glycogen' and 'glucagon breaks down glycogen to glucose' as if they were enzymes.

3 Glucagon circulates in the bloodstream and binds to glucagon receptors on liver cells. The receptor interacts with adenylyl cyclase to increase the concentration of cyclic AMP inside liver cells in the same way as the interaction with adrenaline (Figure 26).

4 Glucagon stimulates liver cells to:
  - activate the enzyme glycogen phosphorylase, which converts glycogen into glucose — this process is called **glycogenolysis** (literally, 'splitting glycogen')
  - convert fat and protein into intermediate metabolites that are converted into glucose — this process is called **gluconeogenesis** (literally, 'making new glucose').

5 Glucose accumulates inside liver cells and diffuses out into the blood, so the concentration of glucose in the blood increases.

Physiologists refer to 'the fasting state' when glucagon secretion takes place. This does not mean that a person is starving, just a period of time between absorbing and assimilating one meal and the next. You are in a fasting state when you are asleep at night; people who have fasting blood tests are told not to eat or drink for 8–10 hours beforehand.

Figure 30 shows how the release of insulin from β-cells is controlled.

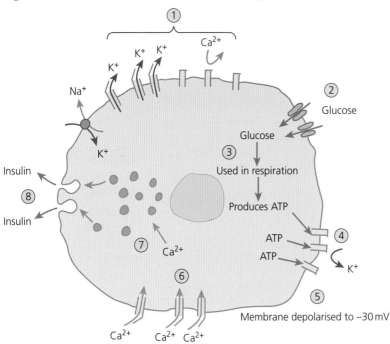

Figure 30 The release of insulin from β-cells in islet tissue in the pancreas

1 In the normal range of blood glucose concentration:
  - membrane potential is at −70 mV; ATP-gated potassium ion channel proteins are open and some potassium ions diffuse out of the cell
  - voltage-gated calcium ion channels are shut; the membrane is impermeable to calcium ions

2 Blood glucose concentration rises as molecules from a meal are absorbed; glucose diffuses into the cell through specific channel proteins (GLUT2).

Knowledge check 17

State how the effect of insulin on liver cells differs from the effect of glucagon.

Knowledge check 18

Glucagon and glycogen are both important in the regulation of blood glucose concentration. Make a table to show five ways in which glucagon differs from glycogen.

3 Glucose is phosphorylated to glucose 6-phosphate in the first stage of respiration (glycolysis) (p. 66), so maintaining the concentration gradient for glucose to continue entering the cell.

4 ATP is produced in respiration; ATP-gated potassium ion channels are sensitive to the increasing concentration of ATP and close.

5 Potassium ions accumulate in the cell thanks to the sodium–potassium ion pump and so the positive charge *inside the cell* increases, causing a depolarisation from −70 mV to +30 mV.

6 Voltage-gated calcium ion channel proteins open in response to the depolarisation and calcium ions diffuse into the cell.

7 Calcium ions stimulate the movement of vesicles towards the membrane (as they do in the synapse).

8 Insulin is released by exocytosis and enters the blood.

Less glucose diffuses into the cell when the glucose concentration falls below the set point. Less respiration occurs and the decrease in ATP concentration causes the ATP-gated potassium ion channel proteins to open.

Diabetes

Insulin is the only hormone that stimulates a decrease in the blood glucose concentration. This means that when something happens to interrupt the secretion of insulin, or its detection by target cells, there is nothing to take over its role. Diabetes mellitus is the disease caused by lack of insulin or an inability of the body to respond to insulin. Diabetes means excessive production of urine and mellitus means sweet. Doctors used to diagnose the disease by tasting the patient's urine.

There are two types of diabetes: type 1 (insulin-dependent) and type 2 (non insulin-dependent). Type 1 diabetes is caused by an inability to secrete insulin, possibly due to destruction of β-cells by the immune system (an autoimmune response). This usually starts at a young age. Type 2 diabetes is an inability of cells to respond to insulin and may happen because there are few receptors on the cell surface membranes of target cells. This form of diabetes is often associated with obesity, a high sugar diet, the inheritance of the alleles of certain genes and ethnicity (people from some ethnic groups are more at risk of diabetes than others).

There is no cure for diabetes. Type 2 diabetes is generally controlled by diet and exercise. Type 1 diabetes is treated by injection of insulin. For many years, diabetes was treated by regular injections of insulin extracted from animals, such as pigs and cattle, slaughtered for the meat trade. This form of insulin is still available, although most diabetics now receive human insulin prepared from genetically modified cells grown in fermenters.

The amino acid sequence of this form of insulin can be changed to alter the properties of the insulin. These substances are called insulin analogues. They either act faster than animal insulin (useful for taking immediately after a meal) or more slowly over a period of between 8 and 24 hours to give the background blood concentration of insulin.

The problem with treating diabetes with insulin is that it has to be injected. It cannot be taken by mouth because it is a protein and would be digested in the gut by proteases. Islet cell transplantation is one possible cure. Cells taken from the islet

Knowledge check 19

Explain the difference between type 1 diabetes and type 2 diabetes.

Exam tip

It is a good idea to read more about current research on type 1 and type 2 diabetes. You could start with the websites of Diabetes UK and Diabetes Research UK.

tissue of donor pancreases are inserted into the liver where the transplanted β-cells secrete insulin. Few transplant patients have stopped taking insulin altogether, although they often need to take less. However, the beneficial effects have not lasted and patients have needed further transplants.

Stem cells are cells that retain the ability to divide by mitosis and differentiate into specialised cells, such as β-cells. Research on mice suggests that transplanted stem cells can differentiate in the host into β-cells and secrete insulin, although there are problems in getting them to respond to changes in blood glucose concentrations. Rejection by the immune system is a serious problem. Putting donated islet tissue or stem cells inside porous capsules may protect them from rejection by the immune system while allowing them to secrete insulin.

Knowledge check 20

Explain briefly how stem cells may be used to treat diabetes.

Synoptic links

Glucose is transported in solution in the plasma. There is not much glucose in the blood or in cells at any one time — as soon as it enters cells it is converted to glucose phosphate. This ensures a steep concentration gradient for glucose. Glucose phosphate is converted into other substances, including glycogen. The advantage of the highly branched structure of glycogen is that many molecules of glucose can be released from glycogen quickly, which is useful for an animal that may have to respond within a split second and needs extra glucose in its bloodstream to be able to do so.

Insulin and glucagon are proteins. Insulin is composed of two polypeptides joined by disulfide bonds and is a good example of a protein with all four levels of organisation. The destruction of β-cells by the immune system is a form of autoimmunity (see Module 4).

Summary

- Hormones are signalling chemicals secreted into blood by ductless (endocrine) glands. They stimulate target tissues that are primed to make responses.
- The cortex of adrenal glands secretes steroid hormones to regulate aspects of metabolism and ion uptake in the kidney. The medulla secretes adrenaline that affects many target tissues during responses to emergencies.
- Adrenaline is a first messenger that interacts with receptors on the cell surface membrane of target cells. In response, cells make cyclic AMP, which is a second messenger that activates the first enzyme in a cascade of enzymes to amplify the message for a fast response.
- Exocrine cells in the pancreas secrete digestive enzymes into ducts that empty into the duodenum. The islets of Langerhans are groups of endocrine cells scattered throughout the exocrine tissue: α-cells secrete glucagon and β-cells secrete insulin to regulate blood glucose concentration.
- Insulin stimulates liver and muscle cells to store glucose in the form of glycogen, so reducing blood glucose concentration. Glucagon stimulates liver cells to break down glycogen to glucose, so increasing blood glucose concentration.
- Insulin secretion is controlled by ATP-gated potassium ion channels in β-cells. They close in response to increases in blood glucose concentration above the set point. The cells depolarise and voltage-gated calcium ion channels open. Calcium ions enter and stimulate movement of vesicles to release insulin by exocytosis.
- Diabetes mellitus is a disease in which the homeostatic control of blood glucose fails to function. In type 1 diabetes β-cells fail to release insulin. Type 2 begins later in life and is a failure of cells to respond to insulin. Type 1 diabetes is treated with injections of insulin; type 2 by other drugs, careful diet and exercise.
- The production of insulin by GM cells allows a constant supply of human insulin, rather than a supply dependent on the meat trade in providing animal insulin. The transplant of stem cells that differentiate into β-cells may be a long-term treatment or even a cure.

Plant and animal responses

Plant responses

Key concepts you must understand

In plant communication systems there are several elements, including:

- receptors to detect stimuli (changes in the environment)
- plant hormones for coordination
- effectors to bring about change

A seed germinates. The embryo inside grows to form a shoot system and a root system. A seed can be orientated in any direction in the soil, so it needs a mechanism to detect gravity so that roots grow downwards and shoots grow upwards. Shoot systems need to respond to the direction of light so that leaves expose the maximum surface area to the light.

Plants respond to stimuli such as gravity, direction of light, grazing by herbivores, water stress, changes in day length and temperature. This is to avoid the **abiotic stress** caused by adverse environmental conditions.

Auxins, gibberellins, cytokinins, abscisic acid (ABA) and ethene are plant hormones (also known as plant growth regulators or PGRs). There are natural and synthetic examples of each type.

Key facts you must know

Plant defences against disease are covered in Module 4. Plants also have defences against being eaten by herbivores. These include:

- release of **pheromones — ethene**
- folding of leaves in response to touch — the sensitive plant, *Mimosa pudica*, which folds its leaves when touched, is an example
- growing physical defences, such as spines and thorns
- producing chemical defences, such as tannins and alkaloids, that poison herbivores or make leaves, roots and stems unpalatable

Deciduous plants lose their leaves at the beginning of a very dry or very cold period when water is not readily available because there is little rainfall or water is frozen in the ground. As with responses to gravity and light, leaf abscission is controlled by hormones.

Tropisms

Directional plant growth responses are called **tropisms**. Shoots grow towards light (positive phototropism) and away from gravity (negative gravitropism).

Phototropins are cell surface membrane proteins that are phosphorylated when illuminated by blue light (see Question 7, p. 81). In the phototropic response they act as receptors that cause more of the auxin indole acetic acid (IAA) to accumulate on the shaded side of the shoot. Auxin stimulates cells to pump hydrogen ions into the

Abiotic stress The negative influences on organisms caused by environmental factors such as hot and cold temperatures, drought, flooding, wind and fire.

> **Exam tip**
>
> Remember from Module 4 that plant pathogens are viruses, bacteria and fungi. Do not confuse the defences against pathogens with defences against herbivory.

Pheromone A chemical released by an organism into the environment as a signal to other members of the same species.

Ethene A gas released by plants that acts as a pheromone to stimulate developmental changes, such as fruit ripening.

cell wall. This acidification loosens the bonds between cellulose microfibrils and the surrounding matrix. Turgor pressure causes the wall to stretch more lengthways so these cells elongate, causing bending towards the light (Figure 31).

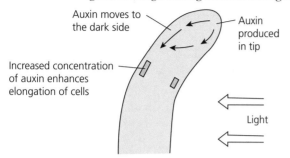

Figure 31 Auxin (IAA) coordinates the positive phototropic response in a shoot

IAA also controls the response to gravity. Cells detect the stimulus because starch grains move to the lowest part of each cell. The effect of this movement is to influence the transport of IAA so that most elongation occurs on the *upper* side of horizontally growing *roots*, so growth is downwards to give a positive gravitropic response, and on the *lower* side of horizontally growing *shoots* to give a negative gravitropic response. This happens because root and shoot cells respond differently to the same concentration of IAA.

Plants can respond to chemicals. When pollen grains start to grow they produce tubes. These grow in response to sources of chemicals that are emitted by the part of the flower that holds the female gamete. Pollen tubes deliver the male gamete so that fertilisation occurs.

Table 3 summarises the roles of the hormones that plants produce. Some of the commercial uses of these hormones are shown in Table 4.

Table 3 The roles of some plant hormones

Plant hormone (with example)	Role
Auxins (IAA)	Stimulate elongation growth Coordinate phototropisms and gravitropisms Control apical dominance: IAA released from the apical bud inhibits growth of lateral buds to form lateral shoots (see Question 3, p. 99) Inhibit abscission of leaves in deciduous plants
Gibberellins (GA₃)	Control stem elongation Stimulate protein synthesis during seed germination
Ethene	Promotes fruit ripening Promotes leaf and fruit abscission
Abscisic acid (ABA)	Controls many responses to abiotic stress Stimulates stomatal closure in dry conditions Promotes bud and seed dormancy
Cytokinins (kinetin)	Promote cell division

Knowledge check 21

Name the response shown by pollen tubes.

Table 4 Commercial uses of some plant hormones

Plant hormone	Example	Commercial use
Auxin	NAA	Rooting compound — stimulates root growth on stem cuttings; used for reducing the number of young fruit on trees (thinning of fruit)
	2,4-D	Weed killer — stimulates rapid elongation growth to kill broad-leaved weeds in cereal crops
Gibberellin	GA_3	Promotes the growth of some fruit crops (e.g. grapes); increases yield of sugar in sugarcane; stimulates germination of barley in malting (start of the brewing process)
Cytokinin	Benzyladenine	Improves fruit quality; prolongs storage life of green vegetables, such as asparagus, broccoli and celery
	Kinetin	Used in tissue culture as it stimulates cell division in buds to give shoot growth

Synoptic links

A synoptic question may ask you to compare hormones in flowering plants with animal hormones. To prepare for such a question you should make a table to compare the two groups of hormones using the following features: sites of synthesis; method of transport; modes of action; and effects (long- and short-term).

Animal responses

Key concepts you must understand

Animals must respond to their environment to obtain food and water, escape from predators, move to suitable conditions and shelter from adverse conditions. They have a sensory system to detect stimuli and two communication systems — the nervous system and the hormonal system. The muscular and skeletal systems act as effectors to bring about responses.

Key facts you must know

The **central nervous system** (CNS) consists of the brain and spinal cord. All the nerves that spread out through the body from the CNS form the **peripheral nervous system** (PNS). **Cranial nerves** are attached to the brain; **spinal nerves** are attached to the spinal cord (see Figure 32a). The neurones within these two systems have specific functions: some are sensory neurones and some are motor neurones (see p. 26). Sensory neurones that send information to the CNS about changes in the external environment and motor neurones that innervate skeletal muscle that moves the body form the **somatic nervous system**. Neurones that control involuntary actions form the **visceral nervous system** (see Figure 32b).

Central nervous system (CNS) The brain and spinal cord.

Peripheral nervous system (PNS) The nerves that originate from the central nervous system and pass throughout the body.

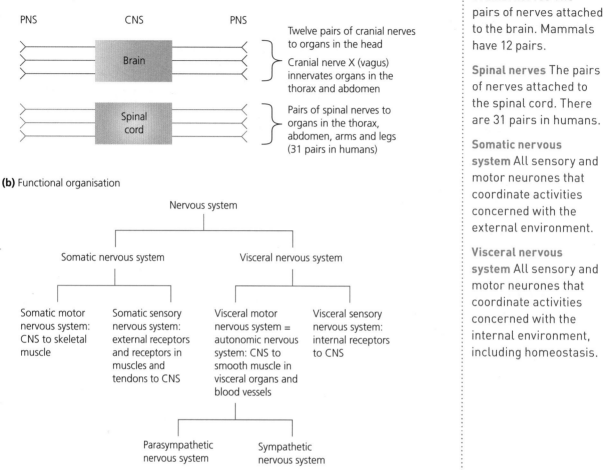

(a) Structural organisation

PNS   CNS   PNS

Brain

Twelve pairs of cranial nerves to organs in the head

Cranial nerve X (vagus) innervates organs in the thorax and abdomen

Spinal cord

Pairs of spinal nerves to organs in the thorax, abdomen, arms and legs (31 pairs in humans)

(b) Functional organisation

Nervous system

Somatic nervous system

Visceral nervous system

Somatic motor nervous system: CNS to skeletal muscle

Somatic sensory nervous system: external receptors and receptors in muscles and tendons to CNS

Visceral motor nervous system = autonomic nervous system: CNS to smooth muscle in visceral organs and blood vessels

Visceral sensory nervous system: internal receptors to CNS

Parasympathetic nervous system

Sympathetic nervous system

Figure 32 The structural and functional organisation of the nervous system in mammals

Cranial nerves The pairs of nerves attached to the brain. Mammals have 12 pairs.

Spinal nerves The pairs of nerves attached to the spinal cord. There are 31 pairs in humans.

Somatic nervous system All sensory and motor neurones that coordinate activities concerned with the external environment.

Visceral nervous system All sensory and motor neurones that coordinate activities concerned with the internal environment, including homeostasis.

Table 5 shows the three types of muscle tissue in vertebrates. The names used in these guides is given first, followed by other names in brackets.

Table 5 Types of muscle

Type of muscle	Cell structure	Distribution	Innervation
Smooth (involuntary, visceral, non-striped, non-striated)	Uninucleate cells; myofibrils not organised	Tubular organs of the viscera such as the gut (oesophagus to anus), airways, Fallopian tubes, uterus, arteries, arterioles and veins	Visceral sensory Autonomic: parasympathetic, sympathetic
Skeletal (voluntary, striated, striped)	Multinucleate (many nuclei in a mass of sarcoplasm); myofibrils organised in parallel bundles	Muscles attached to the skeleton by tendons	Somatic sensory Somatic motor
Cardiac	Uninucleate cells joined by intercalated discs; myofibrils similar to striated	Heart	Visceral sensory Autonomic: parasympathetic, sympathetic

Smooth muscle and cardiac muscle are composed of individual cells. The biceps and triceps in the arm and the quadriceps femoris in the leg are examples of muscles composed of skeletal muscle fibres, which are not composed of individual cells. Each muscle 'cell' or fibre is a syncytium with many nuclei in one mass of sarcoplasm (muscle cytoplasm).

The autonomic nervous system

The autonomic nervous system (ANS) consists of all the motor neurones that innervate organs that are not usually under conscious control. The neurones innervate smooth muscle in the viscera and cardiac muscle in the heart, which receives dual innervation by:

- cardiac accelerator neurones that increase heart rate
- cardiac decelerator neurones that decrease heart rate

The accelerator neurones are part of the **sympathetic nervous system**, which tones up the body in response to stressful situations. This system works in conjunction with the catecholamine hormones adrenaline and noradrenaline. The decelerator neurones are part of the **parasympathetic nervous system**, which tones down the body and helps to conserve resources.

In the **somatic motor system** the cell bodies of neurones are in the brain or spinal cord and impulses pass directly to the muscles (see the reflex arc in Figure 35(b) for an example). Neurones in the autonomic nervous system are different: there are two neurones in series between the CNS and each effector. The synapses between the neurones are in swellings known as ganglia. In the parasympathetic system the ganglia are in or on the innervated organs. In the sympathetic system they are close to the CNS so that impulses pass throughout the body to coordinate responses to danger.

You may read that the two systems act antagonistically. This is true of the control of the heart, but it is not true for all the aspects of physiology (Table 6).

Autonomic nervous system (ANS) The visceral motor nervous system (sometimes used for the whole of the visceral nervous system).

Knowledge check 22

Define the term *antagonistic*. Name two hormones that function antagonistically.

Table 6 Some of the functions of the autonomic nervous system

Organ/tissue	Sympathetic innervation	Parasympathetic innervation
Eye: iris radial muscle	Contracts to dilate pupil	Does not innervate
Eye: iris circular muscle	Does not innervate	Contracts to constrict pupil
Adrenal medulla	Secretion of adrenaline	Does not innervate
Salivary glands	Small quantities of viscous saliva	Large quantities of dilute secretion rich in amylase
Alimentary canal	Inhibits, for example stimulates arterioles to contract (vasoconstriction), diverting blood elsewhere	Activates, for example stimulates muscle contraction and secretion from gastric glands and pancreas
Heart	Increases heart rate	Decreases heart rate
Whole body	Coordinates response to stress, including mobilising resources	Coordinates conservation of resources

Nervous and hormonal control

The nervous and hormonal systems act together to coordinate activities in the body. An example is the control of heart rate.

The heart is myogenic in that the stimulus for contraction originates in the heart itself, in the sinoatrial node (SAN). However, the rate at which the SAN works is influenced by the accelerator nerves, which are adrenergic in that they release noradrenaline as the neurotransmitter, and the decelerator nerves, which release acetylcholine. The nerves also influence the atrioventricular node (AVN). Figure 33 shows this dual innervation of the heart. (Innervation means the supply of nerves to an organ.) The heart also responds to adrenaline and noradrenaline in the blood.

Exam tip

Questions on the structure and function of the heart from Module 3 may also ask about the control by the nervous and endocrine systems.

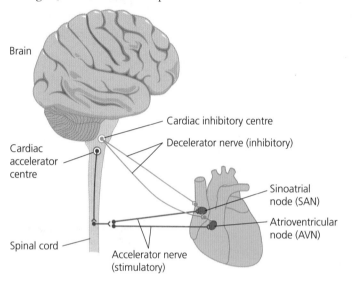

Figure 33 Dual innervation of the heart by accelerator and decelerator nerves

'Fight or flight'

During a dangerous or stressful situation, the nervous and endocrine systems work together to coordinate responses from the whole body. Impulses from the sensory organs travel along sensory neurones in the somatic sensory system. These may stimulate reflexes through the spinal cord and brain. Impulses also travel to the cerebrum in the brain, which makes decisions about how to respond.

Impulses in somatic motor neurones travel to skeletal muscles to coordinate appropriate movements (stand and fight, or turn and flee). Impulses travel along sympathetic neurones in the cardiac accelerator nerves to increase the output of the heart by increasing the heart rate and stroke volume. Sympathetic innervation causes vasoconstriction in the gut and skin and vasodilation in muscles. Blood is thus distributed to where it is needed most to deliver oxygen and glucose. Some sympathetic pre-ganglionic neurones terminate in the adrenal glands, where they stimulate the secretion of adrenaline, which works alongside the sympathetic system, for example by stimulating the liver to convert glycogen to glucose, increasing the blood glucose concentration.

Structure and functions of the human brain

Figure 34 shows views of the human brain and lists the functions of the major parts.

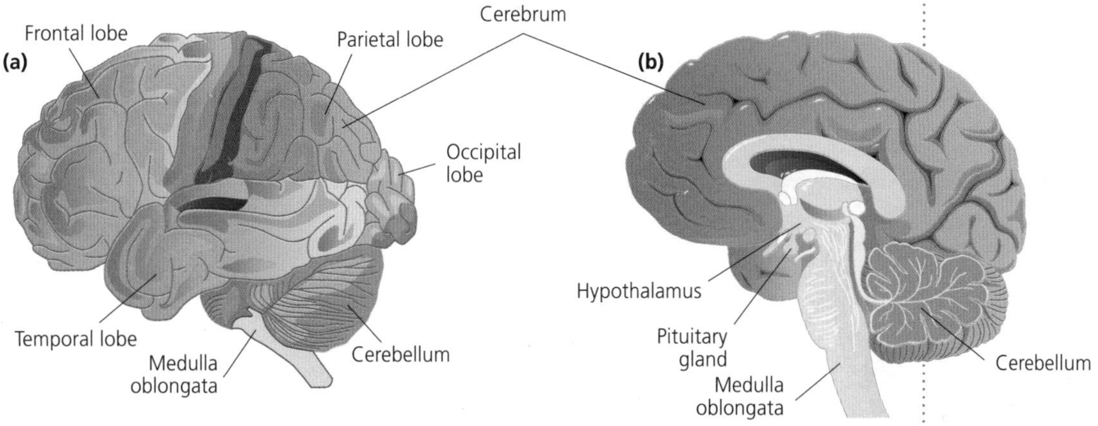

Region of brain	Functions
Cerebrum	Conscious thought Coordination of voluntary activities Memory Association of incoming information with past experience Learning and reasoning Understanding of language Control of speech Interpretation of visual, auditory and other external stimuli
Cerebellum	Interpretation of sensory input from muscles and tendons Coordination of balance, posture and movement Muscle coordination
Hypothalamus	Control of core body temperature, osmoregulation (water potential of blood), reproduction through anterior pituitary, secretion of hormones (e.g. ADH)
Pituitary gland	Anterior pituitary gland produces and releases hormones to control growth, reproduction, metabolism Posterior pituitary gland releases two hormones: ADH (p. 21) and oxytocin (both produced in hypothalamus)
Medulla oblongata	Regulation of autonomic activities (e.g. heart rate, blood pressure, breathing rate and peristalsis)

Figure 34 (a) An external view of the human brain. (b) A vertical section, with the functions of the major parts

The control of muscular movement

The brain, spinal cord and peripheral nervous system (Figure 35a) work together to control the contraction of striated muscle. Neurones do not work in isolation — the simplest way in which they work together is the reflex arc.

Reflex actions

The **knee jerk reflex** is a spinal reflex. Figure 35(b) shows that it is a monosynaptic reflex with two neurones in series. Stretch reflexes like the knee-jerk reflex are involved in maintaining posture and balance and are used during walking. The diagram shows that when the muscle spindle is stimulated by stretching the tendon, impulses travel to the spinal cord. The impulse is then sent across a synapse to the motor neurone that sends impulses to the quadriceps femoris muscle in the leg, which contracts. The circuit is known as a **reflex arc**.

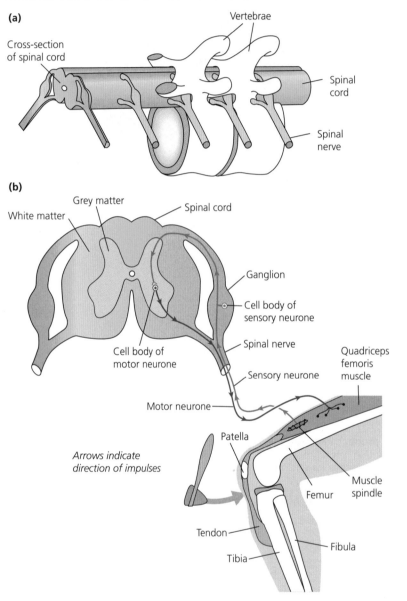

Figure 35 (a) The position of the spinal cord inside the vertebral column and the origin of the spinal nerves. (b) The monosynaptic reflex arc that coordinates the knee-jerk reflex. In many other reflex arcs there is a relay (connector) neurone between the sensory and motor neurones

The **blinking reflex** is a cranial reflex. A variety of stimuli cause this reflex, but they include anything touching the front of the eye. Sensory neurones around the edge of the eye transmit impulses to a centre in the hind brain where they synapse with relay neurones that stimulate motor neurones that terminate in the muscles of the eyelids.

Histology of striated muscle

Figure 36 is a photomicrograph of striated muscle and Figure 37 shows the detailed structure of a striated muscle fibre.

Figure 36 A longitudinal section through some striated muscle fibres. The striations across the fibres and also some nuclei are visible

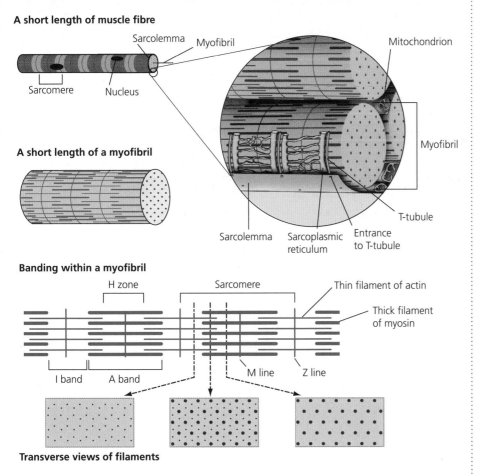

Figure 37 The structure of a muscle fibre

Muscle contraction

Motor neurones terminate at neuromuscular junctions. Action potentials travel down the motor neurones and stimulate calcium ions to enter, causing vesicles of neurotransmitter to fuse with the membrane. Acetylcholine diffuses across the gap and binds to receptors, causing an inflow of sodium ions that depolarise the sarcolemma (an end-plate potential). If above threshold, action potentials are transmitted along the sarcolemma and down into transverse system (T-system) tubules. These membranes have the same channel proteins and pumps that are found in neurones.

The impulse is coupled to the contraction mechanism. Depolarisation of T-system tubules causes calcium channels in the sarcoplasmic reticulum to open and release calcium ions into the sarcoplasm. Calcium ions act as second messengers to stimulate movement of the muscle myofibrils.

When calcium ions bind to troponin they cause a second protein, tropomyosin, to move, exposing binding sites for myosin on thin filaments. Myosin heads move towards the thin filament, bind to actin and swivel, as shown in Figure 38. This swivelling motion is the power stroke that moves the thin filaments closer together, reducing sarcomere length. The combined effect of shortening in all the sarcomeres shortens the length of the myofibril. The combined effect of shortening all the myofibrils in all the muscle cells shortens the length of the muscle. When there is no action potential in the sarcolemma, calcium is pumped back into the sarcoplasmic reticulum and contraction stops.

Exam tip

You should look at good animations of muscle contraction so that you can see the changes that occur to sarcomeres.

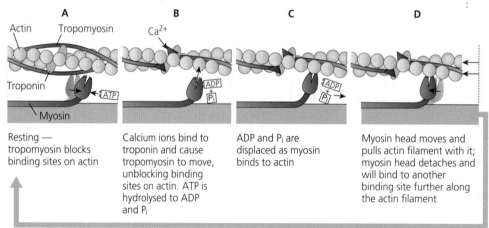

Figure 38 Sliding filaments — changes that occur in a myofibril during contraction

The myosin head is an ATPase, which hydrolyses ATP to ADP and P_i. When it moves during the power stroke ADP and P_i are released. ATP then binds to the myosin head and causes the myosin and actin to separate. The myosin head returns to its original position and is ready to repeat the process.

There is very little ATP readily available within muscle tissue. There is a store of creatine phosphate (CP) within the sarcoplasm where an enzyme transfers phosphate to ADP to resynthesise ATP. There is enough ATP and CP for about 6–8 seconds of exercise. Once this is exhausted, ATP must be synthesised from respiration. At the start of exercise or during short-term strenuous exercise, such as weight lifting and sprinting, the energy comes from respiration of muscle glycogen with the production of lactate, as there is not enough oxygen for aerobic respiration. This is unsustainable in the long term, so during aerobic exercise (e.g. long-distance running) aerobic respiration of carbohydrate and fat occurs to provide the ATP needed for muscle contraction.

Knowledge check 23

List all the ways in which muscle tissue is provided with energy for contraction.

Summary

- Plants respond to changes in their environments to avoid herbivory and abiotic stress (e.g. the effects of drought).
- A tropism is a plant growth response to stimuli such as gravity and the direction of light.
- Plant hormones coordinate activities in plants. Auxins coordinate tropisms and apical dominance; gibberellins control stem elongation. Various hormones coordinate leaf fall in deciduous plants.
- Auxins are used commercially as weedkillers and rooting compounds; gibberellins are used to promote growth of grapes and hasten the germination of barley.
- The human nervous system is divided into the central nervous system (brain and spinal cord) and the peripheral systems (cranial and spinal nerves).
- The autonomic nervous system is composed of motor neurones of the visceral nervous sytem. The parasympathetic system controls conservation of body resources and the sympathetic system works, along with adrenaline, to coordinate responses to danger, stress and physical activity.
- The main regions of the human brain are cerebrum, cerebellum, medulla oblongata and hypothalamus.
- Heart rate is controlled by neurones in the cardiac accelerator and decelerator nerves. Adrenaline also stimulates an increase in heart rate.
- There are three types of muscle tissue in mammals: skeletal (striated), involuntary (smooth) and cardiac. Each striated muscle fibre contains many myofibrils composed of the contractile proteins actin and myosin arranged into thin and thick filaments within each sarcomere.
- Motor neurones stimulate release of acetylcholine at neuromuscular junctions, which stimulates impulses that travel across the sarcolemma and down T-system tubules. These impulses stimulate release of calcium ions into muscle tissue, leading to the movement of myosin heads, which move thin filaments closer together, so shortening myofibrils.
- The energy released by the hydrolysis of ATP is used in muscle contraction. Creatine phosphate helps to maintain the concentration of ATP in muscle tissue.

Energy for biological processes

■ Photosynthesis

Key concepts you must understand

Plants use light energy to convert simple inorganic molecules to complex organic molecules. They are **autotrophs**. Animals and decomposers cannot convert simple molecules to complex molecules. They obtain energy in the form of complex organic compounds and are known as **heterotrophs**. Not all autotrophic organisms use light as a source of energy — some prokaryotes use sources of chemical energy instead.

Photosynthesis provides the complex carbon compounds that are the basis of the metabolism of plants and animals. Carbon dioxide and water are the raw materials and triose (3C) sugar is the product of photosynthesis. This is used by many pathways inside plant cells to make other biochemicals. Oxygen is the by-product of photosynthesis. Some of it is used in aerobic respiration; the rest diffuses out into the surroundings.

Respiration is the metabolic process that transfers energy from complex carbon compounds to ATP, making energy available for cell activities (see p. 63). All organisms carry out respiration. Animals gain their carbon compounds by eating plants or other animals. Decomposers gain carbon compounds from dead plant and animal material.

Key facts you must know

All the stages of photosynthesis occur in chloroplasts. The structure of this organelle and its exchanges with the rest of the cytoplasm are shown in Figure 39.

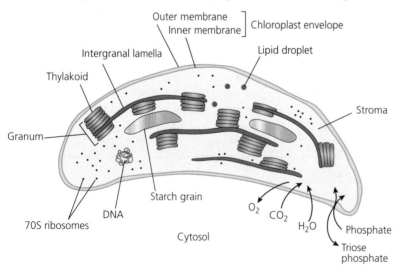

Figure 39 The structure of a chloroplast and the exchanges with the surrounding cytosol in a plant cell

Knowledge check 24

What exchanges occur between the processes of respiration and photosynthesis in plant cells?

Exam tip

Respiration and photosynthesis are not the reverse of each other. Think about energy. Photosynthesis absorbs light. Respiration does *not* emit light. You respire, but you do *not* glow in the dark.

Exam tip

Make a diagram of a spongy mesophyll cell to show the interrelationships between photosynthesis and respiration. Make another diagram of a chloroplast and annotate it with the functions of the different parts of this organelle.

Exam tip

You should be able to recognise the different parts of this organelle in electron micrographs. Have a look at some examples to prepare for labelling the parts of a chloroplast in an exam.

Photosynthetic pigments

A pigment is a coloured compound that absorbs some wavelengths of light and reflects others (see Table 7). The colour we see is the colour that is reflected. A photosynthetic pigment is a coloured compound that absorbs light energy over a range of wavelengths, with a peak at a certain wavelength. Photosynthetic pigments can be separated by paper and thin-layer chromatography.

Table 7 Photosynthetic pigments

Pigment	Colour	Peak absorption/nm	Function in photosynthesis
Chlorophyll *a*	Yellow–green	430, 662	Absorbs red and blue-violet light
Chlorophyll *b*	Blue–green	453, 642	Absorbs red and blue-violet light
β-carotene	Orange	450	Absorb blue-violet light; may protect chlorophylls from damage from light and oxygen
Xanthophylls	Yellow	450–470	

A colorimeter is used to determine the wavelengths that these pigments absorb by exposing solutions of each pigment to light passed through different coloured filters. Figure 40 shows the **absorption spectrum** for three chloroplast pigments. The

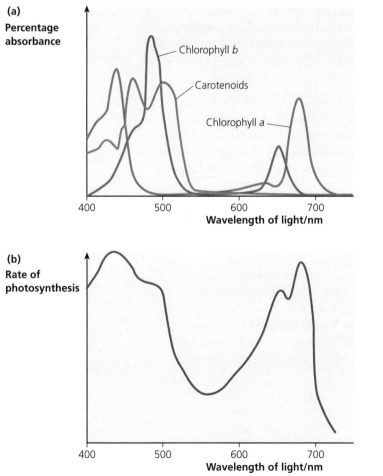

Figure 40 (a) Absorption spectra for photosynthetic pigments. (b) Action spectrum for photosynthesis

> **Exam tip**
>
> You can see that the action spectrum and the absorption spectra coincide, showing that the light absorbed by the pigments is used in photosynthesis.

effectiveness of the pigments in absorbing light for photosynthesis is determined by exposing a plant to different wavelengths of light and measuring the rate of photosynthesis to give an **action spectrum**.

The light-dependent stage

This stage occurs on the thylakoid membranes. Light is absorbed, providing energy to split water. The products of the light-dependent stage are ATP and reduced NADP. The process involves the transport of electrons through a series of compounds linked to the movement of hydrogen ions (protons) across the thylakoid membranes.

Two photosystems, PSI and PSII, are situated in the membranes. These consist of light-harvesting complexes (LHCs) with different pigments. Two types of chlorophyll *a* molecule form the reaction centres of the photosystems. Chlorophyll *a* with peak absorption at 700 nm is the reaction centre of PSI; chlorophyll *a* with peak absorption at 680 nm is the reaction centre of PSII. Pigments in the LHCs absorb photons of light and pass the energy to the reaction centres, P700 and P680.

Light energy excites electrons in P680 in PSII so that they gain enough energy to leave the molecule and pass to a series of protein carrier molecules. As electrons move through the chain (known as the electron transport chain or ETC) they lose energy, which is harnessed to move protons by a form of active transport across the thylakoid membrane into the thylakoid space. The electrons then pass to PSI, where they are re-energised by the absorption of light energy and pass through more carriers to reduce NADP on the stromal side of the membrane.

Proton pumping results in a higher concentration of protons in the thylakoid space than in the stroma. The protons can only leave the thylakoid space through ATP synthase. The diffusion of protons down their concentration gradient through ATP synthase provides energy for the synthesis of ATP from ADP and phosphate ions. The use of light energy to drive the production of ATP is called **photophosphorylation**.

Water provides protons for photophosphorylation, and protons and electrons for the reduction of the coenzyme NADP. In the thylakoids there is a water-splitting enzyme that breaks down water to give hydrogen ions, free electrons and oxygen:

$$2H_2O \rightarrow 4H^+ + 4e^- + O_2$$

The light-dependent stage can be shown as the Z-scheme (Figure 41). This shows the route followed by electrons as they pass from water to NADP in the thylakoid membranes. It is actually a Z rotated 90° to the right and looks more like the letter N. It is a graph in which the vertical axis is the energy level of electrons at different stages of the pathway. In the vertical part of the N, electrons gain energy from light. In the falling part of the N, electrons lose energy, which is used to drive protons into the thylakoid space.

Exam tip

Look at some animations of the light-dependent stage of photosynthesis. A good place to start is John Kyrk's website.

Exam tip

Z-scheme diagrams have a vertical scale that shows each molecule's ability to transfer an electron to the next in the chain. Energy is available to pump protons in the 'downhill' sections.

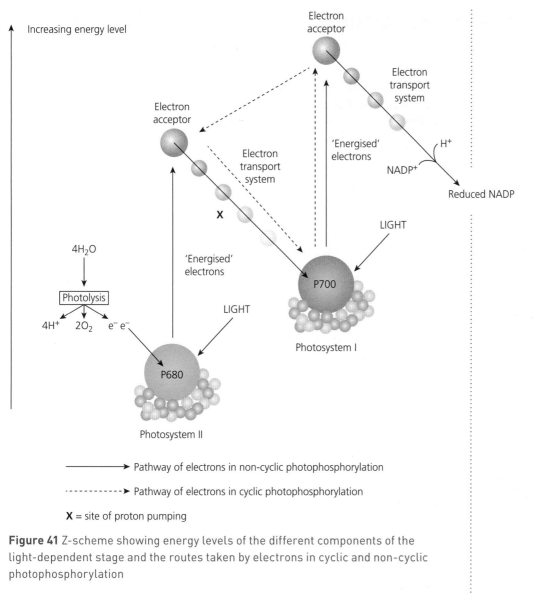

Figure 41 Z-scheme showing energy levels of the different components of the light-dependent stage and the routes taken by electrons in cyclic and non-cyclic photophosphorylation

The sequence of processes in the light-dependent stage is as follows:

- Light energy is absorbed by LHCs.
- Water is split by photolysis to give H⁺ and electrons.
- Oxygen is released.
- Energy is harnessed when electrons flow along chains of electron carriers by active transport of protons (proton pumping). A proton gradient is formed.
- ATP is formed in:
  - cyclic photophosphorylation
  - non-cyclic photophosphorylation
- In non-cyclic photophosphorylation, reduced NADP is formed.

Knowledge check 25

Make a table to compare cyclic with non-cyclic photophosphorylation.

The light-independent stage

The light-independent stage occurs in the stroma and involves the fixation of carbon dioxide.

Carbon dioxide diffuses into leaves through stomata, dissolves in water in the cell wall and diffuses into cells and into chloroplasts. The enzyme in the stroma that fixes carbon dioxide is **ribulose bisphosphate carboxylase/oxygenase**, usually abbreviated to **rubisco**. This enzyme combines carbon dioxide with the 5-carbon acceptor compound **ribulose bisphosphate (RuBP)** to form a 6-carbon compound that immediately breaks down to form two molecules of the 3-carbon compound **glycerate 3-phosphate (GP)**. These molecules are reduced to form **triose phosphate (TP)**, which can be converted into a range of compounds including more RuBP, the carbon dioxide acceptor molecule. Since triose phosphate can be used to regenerate ribulose bisphosphate the reactions form a cycle. The cycle is named after one of the American scientists who discovered it — Melvin Calvin.

Figure 42 shows the Calvin cycle. The three main processes are as follows:

- **carboxylation** — combination of carbon dioxide with RuBP to form GP:

$$CO_2 + RuBP\ (5C) \rightarrow 2 \times GP\ (3C)$$

- **reduction** of GP to TP using reduced NADP from the light-dependent stage
- **regeneration** by **synthesising** the carbon dioxide acceptor, RuBP

Energy in the form of ATP produced by the light-dependent stage drives the Calvin cycle. GP is phosphorylated by ATP to give TP and ribulose phosphate (RP) is phosphorylated to regenerate RuBP.

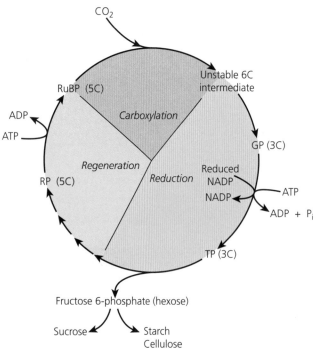

Figure 42 The light-independent stage. ATP and reduced NADP are the products of the light-dependent stage. The product of the light-independent stage is triose phosphate (TP)

Exam tip

You should also follow some good animations of the light-independent stage of photosynthesis. Again, start with John Kyrk's website.

Exam tip

Remember that carbon dioxide is one of the raw materials of photosynthesis.

Exam tip

Rubisco accepts carbon dioxide and oxygen in its active site, hence the inclusion of carboxylase and oxygenase in its name. You only need to know about the carboxylase function.

TP molecules are converted to organic acids for synthesising amino acids by combination with ammonia, using ATP from the light-dependent stage. TP is exported from chloroplasts into the cytosol by an antiport carrier protein in exchange for phosphate ions (see Figure 39). TP molecules may be combined to form hexose phosphates, which are used to synthesise sucrose for transport, starch for storage and cellulose for cell walls. Fatty acids are also made and converted into triglycerides for storage and phospholipids for membranes.

Synoptic links

The light-independent stage is sometimes called the 'dark stage' or 'dark reaction'. This does not mean that it occurs in plants when it is dark. It means that it does not require light directly to function. Instead it requires ATP and reduced NADP. As light intensity decreases in the evening, there is less energy available from the light-dependent stage and the reactions of the Calvin cycle slow down and then stop.

The rate of photosynthesis is dependent on temperature. In Module 2 you studied temperature as one of the factors that influence enzyme action. There are many enzymes involved in the Calvin cycle, so the rate of the light-independent stage is temperature dependent. The effect of temperature as a limiting factor of photosynthesis is described below.

Limiting factors

Light intensity, temperature and carbon dioxide concentration can limit the rate of photosynthesis. When they do so, they are called **limiting factors**. A limiting factor is any environmental factor that restricts a process, such as photosynthesis or growth. When all other factors are favourable, the factor at or closest to its minimum is the factor that is limiting. On hot, sunny days carbon dioxide concentration is the limiting factor for photosynthesis.

Light intensity determines the energy available for the light-dependent stage. As the light intensity increases, more energy is trapped and made available as ATP and reduced NADP for carbon dioxide fixation in the light-independent stage. Light intensity is the limiting factor at dawn and dusk and on cloudy and overcast days.

Carbon dioxide concentration limits the rate of photosynthesis because it is the raw material for the light-independent stage. If there is a limited supply of carbon dioxide, small quantities of triose phosphate are produced and there is a limited demand for ATP. Therefore, the light-dependent stage is slow and little oxygen is produced. The carbon dioxide concentration of the air is 0.04% or 400 parts per million. On warm sunny days, or deep inside the canopy of a forest, or in a crop where competition for the gas is intense, it is likely that carbon dioxide concentration decreases and so it becomes the limiting factor.

The reactions of the light-independent stage are catalysed by enzymes. At low temperatures, the rate of these reactions is slow. As the temperature increases, the reaction rate increases. At high temperatures, enzymes are denatured. At low temperatures, ice crystals form within cells, break cell membranes and so destroy cells.

Exam tip

Carbon dioxide concentration, light intensity and temperature are the three major limiting factors of photosynthesis that you should know. You should be able to describe how to investigate them and analyse and interpret the results of such investigations, as in the graphs in Figure 43.

Like carbon dioxide, water is a raw material for photosynthesis, but it does not act as a limiting factor in the same way since water makes up about 90% of the fresh mass of plants. When water is in short supply, plants often close their stomata to reduce the loss of water in transpiration. The effect of this is to reduce the supply of carbon dioxide by diffusion through stomata.

Exam tip

Figure 43 shows the *net* rate of photosynthesis (also known as the *apparent* rate). This is determined by recording the volume of oxygen released. Some oxygen produced in photosynthesis is used in respiration, so it is impossible to measure the *actual* rate because that oxygen is not released.

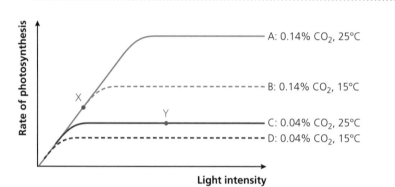

Figure 43 The effect of limiting factors on the rate of photosynthesis

These limiting factors have an effect on the relative concentrations of RuBP, GP and TP in chloroplasts. Look carefully at the graphs in Figure 44.

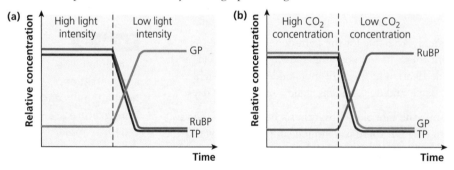

Figure 44 The effects of (a) decreasing light intensity and (b) decreasing carbon dioxide concentration on the relative concentrations of RuBP, GP and TP

In (a) the explanation for the changes is to do with the provision of energy from the light-dependent stage. In (b) it is to do with carbon dioxide fixation in the light-independent stage. When the *light intensity decreases* there is less energy so less ATP and reduced NADP are available. With less energy to drive the cycle there is less TP formed from GP. With little TP and little energy there is little RuBP formed. Fixing carbon dioxide does not require energy so what RuBP is available is used to make GP. When the *carbon dioxide concentration decreases* there is less carbon dioxide to fix. Look at the diagram of the Calvin cycle in Figure 42. If there is less carbon dioxide,

Exam tip

Note the shape of the lines on Figure 43: a slope and a plateau. These are explained in terms of limiting factors. The rate is limited in the slope by the factor that is the independent variable. The rate is limited in the plateau by another factor.

Knowledge check 26

State the limiting factors in regions X and Y in Figure 43.

Knowledge check 27

Many crops are grown in greenhouses (glasshouses). The environmental conditions can be controlled to maximise production. Suggest how *three* such factors may be controlled to maximise rates of photosynthesis.

then RuBP accumulates because it is not being used to fix carbon dioxide. If carbon dioxide is not being fixed, then GP will not be formed, so the concentration decreases, as will that of TP.

Summary

- Autotrophic organisms fix carbon dioxide using energy from the Sun or from simple chemical reactions. Plants use light energy to produce complex organic molecules in photosynthesis. Heterotrophs take in fixed carbon in the form of organic molecules.
- The raw materials for photosynthesis are the products of aerobic respiration, but the two processes are not the reverse of each other.
- Photosynthesis provides triose phosphate, which is used to make organic compounds in plants, including carbohydrates and fats that are respired to provide ATP.
- Photosynthesis is a two-stage process that occurs entirely within chloroplasts.
- The light-dependent stage occurs in the grana. Photosynthetic pigments absorb light energy, transferring it to chlorophyll a molecules at the centre of photosystems I and II (PSI and PSII). Electrons emitted by these molecules travel through the electron transport system with pumping of protons into thylakoids and the formation of reduced NADP. These protons diffuse through ATP synthase, which produces ATP. In cyclic photophosphorylation, electrons flow from and return to PSII with formation of ATP; in non-cyclic photophosphorylation, electrons from the photolysis of water reduce NADP with the generation of ATP.
- Oxygen is a by-product of photolysis and is used in aerobic respiration; the excess diffuses to the surroundings.
- The light-independent stage occurs in the stroma. ATP and reduced NADP drive the reactions of the Calvin cycle. The enzyme rubisco catalyses the fixation of carbon dioxide as glycerate 3-phosphate (GP). Other steps in the cycle are triose phosphate (TP) formation and the regeneration of the carbon acceptor RuBP.
- Triose phosphate is used to make carbohydrates (e.g. sucrose, starch and cellulose), lipids and amino acids. Most is recycled to RuBP.
- Limiting factors for the rate of photosynthesis are light intensity, carbon dioxide concentration and temperature. Changes in all three affect the concentrations of GP, RuBP and TP in the stroma.
- The effects of limiting factors on the rate of photosynthesis are investigated by determining the rate of oxygen production directly by measuring the volume of gas collected from aquatic plants. Most of the gas is oxygen.

Respiration

Key concepts you must understand

Plants, animals and microorganisms respire in order to release energy from carbon compounds and make it available for a variety of functions:

- biosynthesis, for example protein, carbohydrate, fat and nucleic acid synthesis
- active transport, for example sodium–potassium ion pumps
- movement, for example: muscle contraction to move part of the body or the whole body; movement of chromosomes during mitosis and meiosis; movement of phagocytic vacuoles and secretory vesicles; movement of cilia and flagella
- maintenance of body temperature in endotherms
- raising the energy level of glucose at the start of glycolysis (p. 65)

Exam tip

Respiration is a chemical process catalysed by enzymes that transfer energy from carbon compounds to ATP. It occurs in all living cells. Do not confuse respiration with breathing and gas exchange, which are physical processes (Module 2).

Energy is available in compounds, such as carbohydrates, fats and proteins, that on oxidation release energy. When you burn foods in oxygen to find out how much energy they provide, all the energy is released at once. This generates so much heat that, if it happened in cells, proteins would be denatured. Enzymes catalyse reactions in which small changes occur. To release all the energy from a compound such as glucose requires many enzyme-catalysed reactions. Some of these transfer energy directly to ATP. In others, energy is transferred indirectly in the reduction of the coenzymes NAD and FAD.

Key facts you must know

ATP

Adenosine triphosphate (ATP) is the universal energy currency. It is a compound that transfers energy within cells. Revise its structure as a phosphorylated nucleotide from Module 2.

The human body has about 75 g of ATP at any one time. It is not stored and is not transported between cells; it is made as and when it is required. The molecules are small and water soluble, so ATP moves around easily within cells. It is a molecule that many enzymes 'recognise' as it fits into active sites and acts as a coenzyme in many reactions by transferring phosphate groups. The turnover of ATP each day is huge, as each molecule is used and recycled continually.

When ATP is hydrolysed, the terminal bond is broken with the transfer of energy:

$$ATP + H_2O \rightarrow ADP + P_i + 30.5\,kJ\,mol^{-1}$$

This reaction is always coupled with another reaction so that energy is transferred. ATP is good at transferring phosphate groups and transferring energy, which is why it is often called the 'energy currency' of cells.

We have already seen that ATP is produced in photophosphorylation. During respiration, ATP is produced in two processes:

- substrate-level phosphorylation, where transfer of phosphate to ADP occurs in the active site of some enzymes in the cytosol and mitochondria
- oxidative phosphorylation, where ATP synthase catalyses formation of ATP in the membranes of mitochondria

Aerobic respiration

Respiration involves the transfer of energy from organic molecules, such as carbohydrates, fats and proteins, to ATP.

Respiration depends on coenzymes. Coenzyme A takes part in a number of reactions and passes 2-carbon fragments into the Krebs cycle. During respiration, there are a number of dehydrogenation reactions in which hydrogen atoms are released from compounds to reduce the coenzyme NAD, which is mobile and transfers hydrogen atoms to the internal membranes of mitochondria. NAD is a hydrogen carrier.

Aerobic respiration of glucose occurs in four stages in cells:

1 Glycolysis — in the cytosol

2 Link reaction — in the matrix of mitochondria

> **Exam tip**
>
> Look at animations of the stages of respiration. A good place to start is John Kyrk's website.

> **Knowledge check 28**
>
> State the precise locations in a plant cell where ATP is made.

3 Krebs cycle — in the matrix of mitochondria

4 Oxidative phosphorylation — across the inner membrane of mitochondria

Aerobic respiration is summarised in Figure 45.

Exam tip

To help your revision, make a large outline diagram of respiration. As you study this guide and other books and websites you can add more details. Do not worry about the names of all the intermediate compounds you find, but concentrate on the principles that are given here.

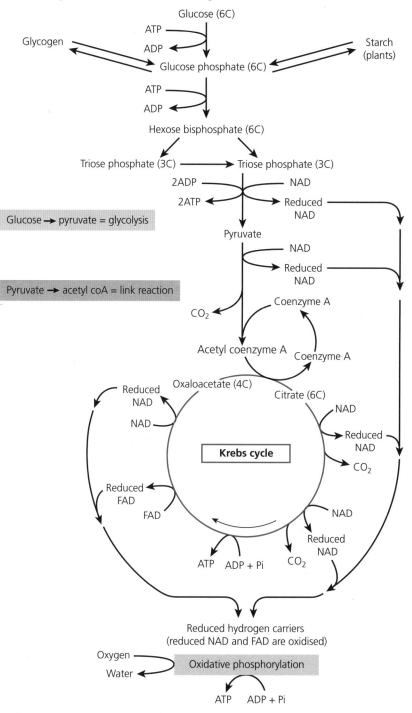

Figure 45 Summary of aerobic respiration

In the first part of glycolysis, glucose is converted into hexose (fructose) bisphosphate (a 6-carbon compound), which is broken down into two molecules of triose phosphate (a 3-carbon compound). The reactions that follow occur *twice* for each molecule of glucose. Remember this when studying the yield of ATP on p. 71.

Glycolysis

Glycolysis literally means the 'splitting of glucose'. The first reaction involves phosphorylating glucose to give glucose phosphate, which is then phosphorylated to hexose (fructose) bisphosphate. (Respiration may begin with glycogen or starch, in which case phosphorylase enzymes first break down these molecules to glucose phosphate.) The next step is the conversion of each hexose bisphosphate molecule into two molecules of triose phosphate, which are then oxidised to pyruvate. During this step a dehydrogenation reaction occurs in which NAD is reduced. Energy from glucose is transferred to NAD and is later transferred to ATP. In the last reactions of glycolysis, ATP is synthesised by substrate-level phosphorylation.

The products per molecule of glucose are:
- 2 × reduced NAD
- 4 × ATP (two ATP were used at the start, so the net gain is 2 ATP)
- 2 × pyruvate

The end product of glycolysis is pyruvate. This is energy rich and is respired in mitochondria if oxygen is present. Pyruvate molecules enter mitochondria through symport channel proteins in the inner membrane. This uptake is driven by the proton gradient across this membrane (p. 68). Pyruvate molecules enter a reaction that links glycolysis to the Krebs cycle.

Exam tip

Note the use of ATP at the start of glycolysis. You may be asked why the *net* production of ATP per glucose molecule in glycolysis is only 2.

Structure of the mitochondrion

Figure 46 shows the structure of a mitochondrion and its exchanges with the cytosol. The outer membrane of mitochondria is freely permeable to many substances. The inner membrane is much more selective, controlling what passes in and out.

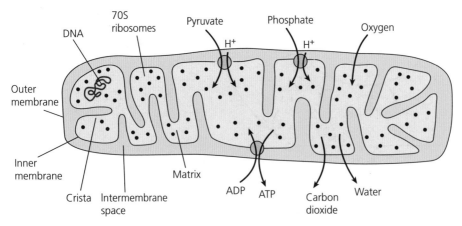

Figure 46 Structure of the mitochondrion and some of its exchanges with the surrounding cytosol

The link reaction

Pyruvate is a 3-carbon compound. During the link reaction (Figure 45) it is converted into the 2-carbon acetyl (ethanoyl) group (sometimes known as 'active acetate'), which combines with coenzyme A to make acetyl coenzyme A. This conversion involves decarboxylation (removal of carbon dioxide) and dehydrogenation (removal of hydrogen, i.e. oxidation). The hydrogen atoms are accepted by NAD, and reduced NAD is formed.

Pyruvate dehydrogenase is a complex made of a number of polypeptides for the three intermediate reactions that form the link between glycolysis and the Krebs cycle. Acetyl coenzyme A transfers the 2-carbon fragment (acetyl group) into the next stage, which is the Krebs cycle (named after Sir Hans Krebs, who discovered it).

The products of the link reaction per molecule of glucose are:
- 2 × reduced NAD
- 2 × CO_2
- 2 × acetyl (2C) coenzyme A

The Krebs cycle

The acetyl group is passed from coenzyme A to oxaloacetate to form citrate. Enzymes in the mitochondrial matrix catalyse reactions to regenerate oxaloacetate, which is the acceptor substance for the acetyl group. During the Krebs cycle, decarboxylation and dehydrogenation reactions occur. Carbon dioxide diffuses out of mitochondria; hydrogen atoms reduce the coenzymes NAD and FAD.

Following the Krebs cycle round (Figure 45), the processes are as follows:
- reaction between acetyl coenzyme A and oxaloacetate — coenzyme A delivers a 2C fragment (acetyl group) to form citrate (6C)
- decarboxylation (× 2) — removal of carbon dioxide
- dehydrogenation (× 4):
  - removal of hydrogen from intermediate substances, which are oxidised
  - reduction of NAD (× 3)
  - reduction of FAD (× 1)
- ATP synthesis (× 1) — substrate-level phosphorylation
- regeneration of oxaloacetate (4C compound)

Products per molecule of glucose (remember, there are two 'turns' of the cycle per molecule of glucose):
- 4 × carbon dioxide
- 2 × reduced FAD
- 6 × reduced NAD
- 2 × ATP

Oxidative phosphorylation

Glycolysis, the link reaction and Krebs cycle generate reduced NAD and reduced FAD. There is only a small quantity of these coenzymes in a cell, so in order for the reactions to continue it is necessary for these reduced coenzymes to be oxidised.

Exam tip

The enzyme complex that catalyses the link reaction is known as pyruvate dehydrogenase. Although it decarboxylates pyruvate, it is not known as a decarboxylase. Pyruvate decarboxylase is a different enzyme that catalyses a different reaction.

Exam tip

You may come across the names of intermediates in the Krebs cycle. The two you should know are oxaloacetate and citrate. The enzymes are omitted from most diagrams of the cycle. One of these is often used as an example of competitive inhibition.

This oxidation transfers energy to ATP. This occurs during oxidative phosphorylation, which is the fourth stage of aerobic respiration (Figure 47).

Reduced coenzymes provide hydrogen atoms. Reduced NAD and FAD are oxidised to provide protons and electrons. The electrons pass along the electron transport chain. The energy released is used to pump protons from the mitochondrial matrix into the intermembrane space. This creates a proton gradient, which is harnessed to make ATP. The protons diffuse from the intermembrane space to the matrix through ATP synthase, which generates ATP. The energy that drives the whole process is the energy from the respiratory substrates that were oxidised during the previous three stages.

Electrons combine with oxygen to form water. This reaction is catalysed by cytochrome oxidase. Oxygen is the final electron acceptor. The water produced in the reaction is sometimes known as metabolic water.

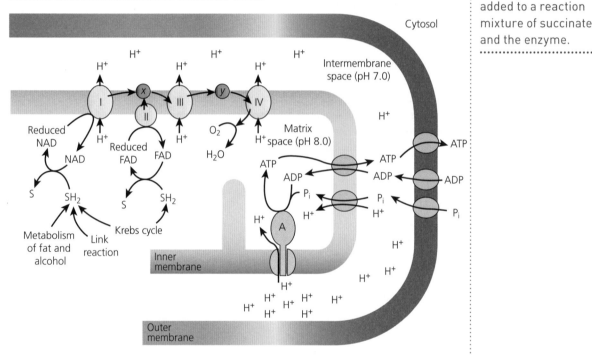

SH_2 = intermediate substances in the link reaction, Krebs cycle and metabolism of fat and alcohol. These substances are dehydrogenated to form reduced hydrogen carriers (reduced FAD and reduced NAD), which are oxidised by complexes I and II

I, II, III and IV are protein complexes *x* and *y* are electron carrier molecules

Complexes I, III and IV pump protons; complex II passes electrons from reduced FAD to complex III via electron carrier *x*. Complex II is not a proton pump

A is a molecule of ATP synthase

Figure 47 Oxidative phosphorylation

The processes in oxidative phosphorylation are as follows:

■ Oxidation of coenzymes NAD and FAD — coenzymes are recycled for use in metabolic pathways in the matrix of the mitochondrion: the link reaction, the Krebs cycle and the metabolism of fats and alcohol.

Knowledge check 29

Succinate dehydrogenase catalyses the dehydrogenation of succinate in the Krebs cycle. Malonate has a molecular structure similar to succinate. Explain why the rate of the reaction slows when malonate is added to a reaction mixture of succinate and the enzyme.

- Electrons flow along the electron transport chain — this provides energy for the active transport of protons.
- Protons are pumped from the matrix into the intermembrane space through three protein complexes (labelled I, III and IV on Figure 47).
- Protons diffuse through ATP synthase — ATP synthase catalyses ATP synthesis:

$$ADP + P_i \rightarrow ATP$$

- The final electron acceptor is oxygen, which is reduced to water.

The products of oxidative phosphorylation are:
- ATP
- water
- NAD and FAD (oxidised forms)

Chemiosmosis

Chemiosmosis is the process by which protons are pumped to create a gradient and the energy in that gradient is used to synthesise ATP. It occurs in chloroplasts, mitochondria and bacteria. Protons are pumped through protein complexes that are part of the electron transport chain. The high concentration of protons in the intermembrane space in mitochondria and in the thylakoid space in chloroplasts gives rise to a potential difference across the inner mitochondrial membrane and the thylakoid membrane. This gives an electrochemical gradient across these membranes, which are impermeable to protons except by diffusion through channels in the enzyme ATP synthase.

The proton gradient and membrane potential are a **proton-motive force** that drives ATP synthesis. This electrochemical gradient acts like a battery, as a store of potential energy — in this case, for the synthesis of ATP. In mitochondria, it is recharged continually using energy from oxidised food. In chloroplasts, it is recharged during the day using light energy absorbed by photosynthetic pigments. It cannot function in chloroplasts in the dark.

Respiration without oxygen

If oxygen is not available, then the last stage of oxidative phosphorylation cannot occur because it needs oxygen as the final oxygen acceptor. Reduced NAD and FAD are not oxidised and the Krebs cycle and the link reaction stop. Pyruvate is no longer moved into the mitochondria. Pyruvate could be excreted, but there would have to be a way to recycle the NAD that takes part in glycolysis. Respiration continues in the absence of oxygen to recycle NAD, using pyruvate. Figure 48 shows what happens to the TP and pyruvate produced in glycolysis in mammals and in yeast when no oxygen is available. Remember that in both cases two molecules of ATP were used in earlier reactions (Figure 45). Table 8 compares respiration in yeast and mammals when there is no oxygen available.

Knowledge check 30

State the precise locations of the four stages of aerobic respiration.

Exam tip

Search online for animations of ATP synthase to see how it uses the proton gradient to generate ATP.

Knowledge check 31

State the roles of coenzymes in respiration.

Exam tip

To help your revision, make a large drawing showing a mitochondrion and a chloroplast. Add information about structure and function and show the exchanges that occur between them in plant cells.

(a)

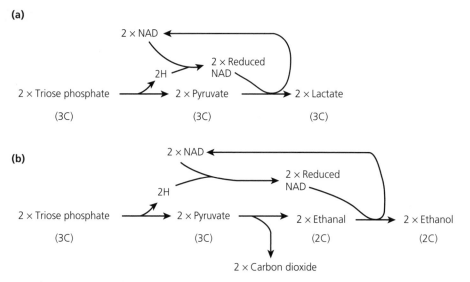

(b)

Figure 48 Fate of pyruvate (a) in mammals and (b) in yeast when there is no oxygen available

Table 8 Respiration in mammals and yeast in the absence of oxygen

Feature	Mammal	Yeast
Decarboxylation (to give carbon dioxide)	No	Yes
Oxidation of reduced NAD	Yes	Yes
Hydrogen acceptor	Pyruvate (3C)	Ethanal (2C)
Products per molecule of glucose	2 × lactate (3C)	2 × ethanol (2C); 2 × CO_2
Net gain of ATP per molecule of glucose	2	2
Reversible reaction(s)	Yes (in liver and heart muscle)	No

Knowledge check 32

Explain the advantage of not excreting lactate.

The type of respiration in mammals is **lactate fermentation**:

$$C_6H_{12}O_6 \rightarrow 2CH_3CHOHCOOH$$

with a net gain of 2ATP per molecule of glucose.

This occurs in muscle and certain other tissues when there is a shortage of oxygen. The advantage of this in muscle is that energy is provided even though there is limited availability of oxygen. The lactate produced diffuses from muscle tissue into the blood. Cardiac muscle in the heart uses lactate in respiration; the rest is recycled to glucose and glycogen in the liver, so energy is not lost. The process in mammals is reversible.

In yeasts and plants, pyruvate is first decarboxylated to ethanal, which acts as a hydrogen acceptor to form ethanol. The type of respiration in yeast (and plants) is **alcohol fermentation**:

$$C_6H_{12}O_6 \rightarrow 2C_2H_5OH + 2CO_2$$

with a net gain of 2ATP per molecule of glucose.

Yeast and plants cannot metabolise ethanol, so this form of respiration is irreversible.

Exam tip

Respiration without oxygen is often called anaerobic respiration, but this is confusing since many bacteria respire in the same way as shown in Figure 45 using substances other than oxygen as their final electron acceptor. This explains the use of the terms lactate fermentation and alcohol fermentation here.

Yield of ATP

Table 9 shows how to calculate the theoretical maximum yield of energy in the form of ATP from aerobic respiration. It is currently thought that during oxidative phosphorylation each reduced NAD molecule provides the energy to make 2.5 ATP molecules and each reduced FAD molecule provides the energy to make 1.5 ATP molecules.

Table 9 The yield of ATP from the complete oxidation of a molecule of glucose in aerobic respiration

Stage of aerobic respiration	Input of ATP (phosphorylation of hexose)	Direct yield of ATP (substrate-level phosphorylation)	Indirect yield of ATP via reduced NAD and reduced FAD
Glycolysis	+2	4	3 *or* 5 (NAD)*
Link reaction		0	5 (NAD)
Krebs cycle		2	15 (NAD); 3 (FAD)
Totals	−2	6	26 *or* 28*

This gives a theoretical maximum yield of 32 − 2 = 30 molecules of ATP (*or 32 depending on how the hydrogens from reduced NAD from glycolysis enter the mitochondria — there are two methods for this, which give rise either to three or to five molecules of ATP).

This theoretical maximum is rarely, if ever, reached for several factors, including the use of the proton gradient to drive the movement of substances into the matrix, for example pyruvate and phosphate ions. Also, protons are lost from mitochondria to the cytosol through the outer membrane.

The transfer of energy to ATP is not 100% efficient, and much energy is 'lost' as heat. Endotherms can retain this heat in the body to maintain a constant body temperature. Ectotherms lose the heat to their surroundings.

Exam tip

The important point to remember here is *not* the calculation in Table 9, but the difference between the net yield in aerobic respiration (about 30 ATP molecules) and the net yield in lactate and alcohol fermentation ('anaerobic respiration') which is 2 ATP molecules.

The 'metabolic funnel'

At A-level we deal with the respiration of glucose, but animals respire glycogen, lots of fat and, in the case of carnivores, quite a lot of protein. Substances like these are **respiratory substrates**. These compounds have to be converted into a form that can be respired, for example fats are hydrolysed to fatty acids and glycerol; fatty acids are then converted into many molecules of 'active acetate' for the Krebs cycle and oxidative phosphorylation, both of which are common to the aerobic respiration of all these substrates.

Amino acids must first be deaminated (p. 14) before they can be respired. An amino acid that has had its amino group removed is an organic acid. Many of these organic acids are similar to those in the Krebs cycle. Therefore, respiring amino acids produces almost as much energy as respiring carbohydrates. The 'metabolic funnel' is the idea that different respiratory substrates are first converted into a form that can enter the Krebs cycle. Glucose, for example, is converted by glycolysis and the link reaction into 'active acetate'.

Table 10 shows the energy values for three respiratory substrates. The energy values of substrates are found by calorimetry — burning substances in oxygen and recording the heat released. The energy content depends on the number of hydrogen atoms per molecule that are available to reduce NAD and FAD. Lipids (fats) have the most. The molecules are composed mainly of carbon and hydrogen so there are many C–H bonds. When these are oxidised, there is the release of much energy per gram.

Table 10 Energy values of respiratory substrates

Respiratory substrate	Energy/kJ g^{-1}
Carbohydrates, e.g. starch, glycogen, glucose, sucrose and lactose	16
Lipids, e.g. triglycerides	39
Proteins	17

Knowledge check 33

Explain why the energy value of lipids is far greater than that for carbohydrates.

Respiratory quotient

The respiratory quotient (RQ) is the ratio between the volume of carbon dioxide produced and the volume of oxygen consumed. The formula to calculate RQ is:

$$RQ = \frac{\text{carbon dioxide produced}}{\text{oxygen consumed}}$$

This is a ratio, so has no units. Values of RQ can be determined from results taken with respirometers.

RQ values are useful as they tell us:

- whether respiration is aerobic or not
- the type of substrate that is respired (e.g. carbohydrate, lipid or protein) or the likely mix of substrates

Table 11 shows the values of RQ and what they tell us.

Table 11 Respiratory quotients

RQ	Explanation
∞	Carbon dioxide is produced, but no oxygen is absorbed
>1	A mixture of aerobic respiration and lactate fermentation/alcohol fermentation
≤1	Respiration is aerobic
1	Carbohydrate is the respiratory substrate
0.9	Protein is the respiratory substrate
0.7	Lipid is the respiratory substrate

It is quite common to find RQ values that are intermediate between these. For example, the RQ for humans is about 0.85, indicating that we respire a mixture of carbohydrate, lipid and protein.

Respirometers

The rates of respiration of yeast, bacteria, small animals, germinating seeds and leaves can be measured in simple respirometers, such as the one in Figure 49.

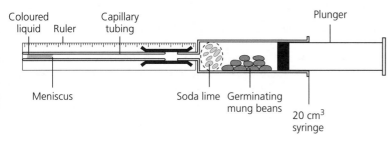

Figure 49 A simple respirometer for measuring the rate of respiration of seeds, small leaves and small invertebrates, such as blowfly larvae

Exam tip

The data in Question 5 on p. 77 were obtained from a simple respirometer like this one.

Soda lime absorbs any carbon dioxide that is produced. As oxygen is absorbed the volume of air in the respirometer decreases and so does the air pressure. The greater air pressure in the environment moves the meniscus to the right. The speed of movement is a measurement of the rate of respiration. The volume of carbon dioxide produced is determined by repeating the procedure under *exactly the same conditions*, but without the soda lime.

- If the meniscus does not move at all it means that the rates of uptake of oxygen and production of carbon dioxide are the same and the RQ = 1.
- If the meniscus moves *away* from the organisms (to the left) then the production of carbon dioxide is greater than the uptake of oxygen (RQ > 1).
- If the meniscus moves *towards* the organisms (to the right) then the uptake of oxygen is greater than the production of carbon dioxide (RQ < 1).

The actual values for rates of uptake of oxygen and production of carbon dioxide can be calculated if the diameter of the capillary tubing is known. If comparative results are required, then they need to be standardised by calculating the rates per unit mass of living material per unit time (e.g. $cm^3 g^{-1} h^{-1}$).

Synoptic links

Temperature is the main environmental factor that influences respiration. This is because all the reactions of respiration are catalysed by enzymes. The body temperature of endotherms is independent of their surroundings, so they have constant rates of respiration and can be active when it is cold.

Summary

- All organisms respire, transferring energy from organic molecules, such as glucose and fat, to ATP, which provides energy for active transport, biosynthesis and movement.
- The four stages of aerobic respiration are: glycolysis (cytosol), link reaction, Krebs cycle (matrix of mitochondrion) and oxidative phosphorylation (cristae).
- Glycolysis begins with the phosphorylation of glucose by 2 × ATP to form hexose bisphosphate. This splits into two molecules of triose phosphate, which are oxidised to pyruvate with the production of 2 × reduced NAD; 4 × ATP are formed by substrate-level phosphorylation.
- In the link reaction, pyruvate is dehydrogenated to form reduced NAD and decarboxylated to form carbon dioxide and an acetyl group (2C).
- Coenzyme A transfers an acetyl group to oxaloacetate (4C) to form citrate (6C). In the Krebs cycle, the 2C compound is decarboxylated to form 2 × CO_2, oxidised to form 3 × NAD and 1 × FAD, and oxaloacetate is regenerated. Substrate-level phosphorylation occurs to produce 1× ATP. The link reaction and Krebs cycle occur twice for each glucose molecule.
- Reduced NAD and reduced FAD are recycled by oxidative phosphorylation. Electrons flow along an electron transport chain with transfer of energy to pump protons from the matrix into the intermembrane space. Protons flow down their gradient through ATP synthase, which phosphorylates ADP to ATP. The final electron acceptor is oxygen, with the formation of water.
- Chemiosmosis is the use of proton pumps to make a proton gradient across membranes; the gradient drives the synthesis of ATP by ATP synthase. The many internal thylakoids in chloroplasts and folded cristae in mitochondria give large surfaces for the many protein complexes.
- Yields of ATP rarely match those expected because respiratory substrates, such as glucose and fat, are not always oxidised completely; energy is used in exchanging substances (e.g. pyruvate) between mitochondria and cytosol; some protons 'leak' through the outer mitochondrial membrane.
- In anaerobic conditions, pyruvate does not enter mitochondria so that much less energy is transferred. In mammals, pyruvate is the final electron and hydrogen acceptor to recycle NAD and form lactate. In yeast and plants, pyruvate is decarboxylated to ethanal, with the formation of carbon dioxide. Ethanal is the hydrogen acceptor and forms ethanol when NAD is recycled. The only ATP produced is during glycolysis, which explains the much lower yield than in aerobic respiration.
- Lipids are the most energy-rich respiratory substrate as they have a higher ratio of hydrogen to carbon and are more highly reduced than carbohydrates and proteins. On oxidation they transfer most energy per gram to ATP.
- The respiratory quotient (RQ) is the ratio between the volume of carbon dioxide produced and volume of oxygen consumed. These volumes are determined for seeds and small invertebrates by using simple respirometers. RQ values indicate the type of respiration and the type of respiratory substrate used.

Questions & Answers

Exam format

At A-level there are three exam papers. Questions in these three papers will be set on any of the topics from Modules 2–6 in the specification. In addition, there will be questions that will test your knowledge and understanding of practical skills from Module 1 and your ability to apply mathematical skills.

Your exams will be as follows:

Paper number	1	2	3
Paper name	Biological processes	Biological diversity	Unified biology
Length of time	2 hours 15 minutes	2 hours 15 minutes	1 hour 30 minutes
Total marks	100	100	70
Types of question	15 multiple-choice questions (1 mark each) and structured questions for 85 marks	15 multiple-choice questions (1 mark each) and structured questions for 85 marks	Structured questions
Synoptic questions	Yes	Yes	The whole paper is synoptic

About this section

The first part of this section contains questions similar in style to those you can expect to find in Paper 1 (Biological processes). The questions in the paper are based on topics from Modules 2–5. Most of the questions on pp. 76–93 are based on topics from Module 5, but some require you to apply your knowledge of Modules 1–4.

The answers to the five multiple-choice questions are on p. 101.

Questions similar to those in Paper 3 (Unified biology) are on pp. 94–100. Paper 3 tests your knowledge of all the Modules (1–6) but concentrates on the skills you have developed during your practical work. Most of the questions on pp. 94–100 are based on topics in Module 5 but they require more knowledge of Modules 1–4 than Paper 1, and some of the questions are set in the context of Module 6.

The limited number of questions in this guide means that it is impossible to cover all the topics and all the question styles, but they should give you an indication of what you can expect in Papers 1 and 3.

As you read through the answers to the questions based on Paper 1, you will find answers from two students. Student A gains full marks for all the questions. This is so that you can see what high-grade answers look like. Student B makes a lot of mistakes — often these are ones that examiners encounter frequently. I will tell you how many marks student B gets for each question. The Paper 3-style questions only have model answers, similar to those of student A.

Examiner's comments

Each question is followed by a brief analysis of what to watch out for when answering the question (icon **ⓔ**). Some student responses are then followed by examiner's comments. These are preceded by the icon **ⓔ** and indicate where credit is due. In the weaker answers, they also point out areas for improvement, specific problems and common errors, such as lack of clarity, weak or non-existent development, irrelevance, misinterpretation of the question and mistaken meanings of terms.

■ Paper 1-style questions: Biological processes

Section A: Multiple-choice questions

Question 1

The table shows the relative plasma concentrations of urea in four blood vessels of the systemic circulatory system.

| | Relative concentrations of urea in blood plasma | | | |
	Hepatic artery	Hepatic vein	Renal artery	Renal vein
A	High	Low	High	Low
B	High	Low	Low	High
C	Low	High	High	Low
D	Low	High	Low	High

Which shows the relative concentrations in these four blood vessels? (1 mark)

Question 2

The speed of conduction of nerve impulses in a giant axon of a squid was determined by varying the distance between a stimulating electrode and a recording electrode. The time for an impulse to travel each distance was plotted on the graph.

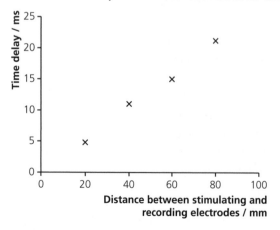

Which is the speed of conduction in the giant axon of the squid?

A $0.4\,mm\,ms^{-1}$

B $4\,m\,s^{-1}$

C $40\,mm\,s^{-1}$

D $4 \times 10^{-4}\,mm\,s^{-1}$ (1 mark)

Question 3

In the autoimmune disease myasthenia gravis antibodies against acetylcholinesterase are produced. What is the effect of these antibodies?

A acetylcholine cannot bind to ligand-gated channel proteins

B acetylcholine is not released from the pre-synaptic membrane

C acetylcholine remains in the synaptic cleft

D sodium ions cannot pass into the post-synaptic neurone   (1 mark)

Question 4

Which plant hormone stimulates the production of amylase in barley grains during germination?

A abscisic acid

B auxin

C cytokinin

D gibberellin   (1 mark)

Question 5

Some crickets of mass 4.5 g were kept in a simple respirometer. Soda lime was placed in the respirometer. The change in volume of gas inside the respirometer was recorded at intervals. The soda lime was then removed from the respirometer and the procedure repeated. The table shows the results.

Time/min	Change in gas volume with soda lime/cm^3	Change in gas volume without soda lime/cm^3
0	0.0	0.0
20	−0.8	−0.2
40	−1.6	−0.4

Which is the carbon dioxide output by the crickets?

A $0.16\,cm^3\,g^{-1}\,h^{-1}$

B $0.24\,cm^3\,g^{-1}\,h^{-1}$

C $0.32\,cm^3\,g^{-1}\,h^{-1}$

D $0.40\,cm^3\,g^{-1}\,h^{-1}$ (1 mark)

Section B: Structured questions

Question 6

(a) Explain the role of receptor cells in coordination in mammals.   (3 marks)

Pacinian corpuscles are skin receptors. The unmyelinated end of a sensory neurone is surrounded by a gelatinous capsule. In an investigation, these receptors were stimulated as shown in Figure 1. Recordings were made with electrodes at A and at B. The results are shown in Figure 2.

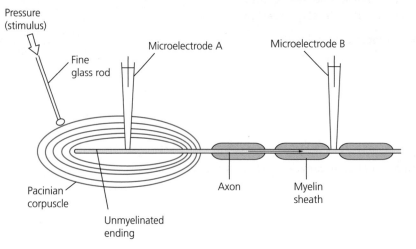

Figure 1

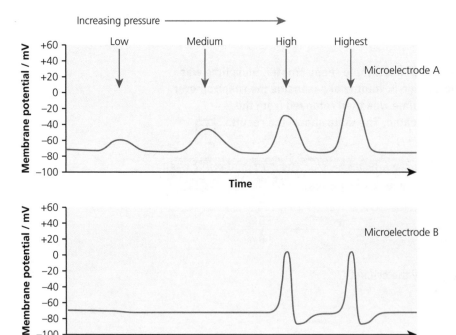

Figure 2

(b) Describe and explain the results shown in Figure 2. (4 marks)

ⓔ Plan answers carefully: use the figures and remember the appropriate terminology to use. 'Membrane potential' on the *y*-axes is a clue.

(c) (i) When the investigation started the sensory neurone was at resting potential. Explain how a resting potential is maintained across the membrane of a neurone. (4 marks)

ⓔ Start by giving the value of the resting potential and then explain that it is due to the distribution of ions across the membrane.

(c) (ii) Explain why the investigators recorded the activity of the sensory neurone at position B. (4 marks)

ⓔ Think about where else they could have put it, i.e. just to the left and right of C.

Student A

(a) Receptor cells detect stimuli from outside and inside the body so that mammals are able to respond to changes in their surroundings and inside their bodies. They are transducers because they convert the energy of the stimulus into the electrical energy of nerve impulses. There are different types of receptor because each is specialised to detect a certain form of energy, for example light (rods and cones) and chemicals (chemoreceptors).

ⓔ **3/3 marks awarded** Student A uses all the terms that are relevant here, especially *transducer*. The question does not ask for examples but it helps to give some specific receptor cells.

Student B

(a) Receptor cells respond to stimuli (changes in the environment) and send impulses along sensory neurones to the CNS.

ⓔ **2/3 marks awarded** Student B has not made the point about transduction.

Student A

(b) Stimulating the corpuscle leads to depolarisation of the unmyelinated ending of the neurone, which shows a graded response to the different intensities of the stimulus. The neurone only fires impulses when the generator potential reaches the threshold of about −40 mV. The neurone follows the all-or-nothing law — it only sends an impulse if threshold is reached. Each action potential in the results from microelectrode B has the same amplitude.

ⓔ **4/4 marks awarded** Student A gives the correct terms to describe the results.

Questions & Answers

Student B

(b) Electrode A records the receptor potential. Electrode B records action potentials going along the sensory neurone. It takes three stimuli to get the neurone to send an impulse.

ⓔ **0/4 marks awarded** Student B refers to the receptor (generator) potential, but has not realised that this must be above threshold before the neurone sends impulses.

Student A

(c) (i) The resting potential of −70 mV is due to the distribution of ions across the axon membrane. It is maintained because Na^+ ions cannot diffuse through the membrane and there is a higher concentration of Na^+ outside the membrane, which makes the outside positively charged. There is a higher concentration of K^+ ions inside the neurone than outside, but inside there are negatively charged proteins and other compounds that cannot pass through the membrane. These anions give a negative charge. The sodium–potassium pump maintains the unequal distribution of Na^+ and K^+ by pumping out three Na^+ for every two K^+ pumped in.

ⓔ **4/4 marks awarded** Student A has written a rather lengthy answer that could have been written more concisely. The important points are:

- the unequal distribution of ions, with a higher concentration of Na^+ outside the membrane than inside and a higher concentration of K^+ inside than outside
- the sodium–potassium ion pump maintains this unequal distribution of ions
- the presence of negatively charged compounds inside the cytoplasm
- the impermeability of the membrane to Na^+
- the voltage-gated channel proteins for Na^+ are closed

Student B

(c) (i) The sodium–potassium pump is responsible for the resting potential by pumping sodium ions out of the neurone and pumping potassium ions into the neurone.

ⓔ **1/4 marks awarded** Student B has made just one of the important points.

Student A

(c) (ii) C is at a node of Ranvier. The sensory neurone is myelinated and the impulse travels in a saltatory way, with action potentials only occurring at nodes.

ⓔ **4/4 marks awarded** This is a concise answer, for full marks.

Student B

(c) (ii) The recording electrode is positioned at C because this is a node of Ranvier where ions flow into and out of the axon. They do not flow in the bits either side because these are covered in myelin and there are few channel proteins there.

e **4/4 marks awarded** Student B has also given a good answer. An excellent answer would combine the points both students have made and state that myelin is an electrical insulator.

Question 7

Shoots are positively phototropic. Membrane proteins known as phototropins are thought to be involved as light receptors in the phototropic response.

The response to unidirectional light was investigated in thale cress, *Arabidopsis thaliana*, by using light of different wavelengths. The effectiveness of the different wavelengths in stimulating a phototropic response in seedlings of thale cress was determined and is shown in Figure 1. Phototropin was extracted from thale cress. The absorption of light of different wavelengths by a sample of phototropin was determined and is shown in Figure 2.

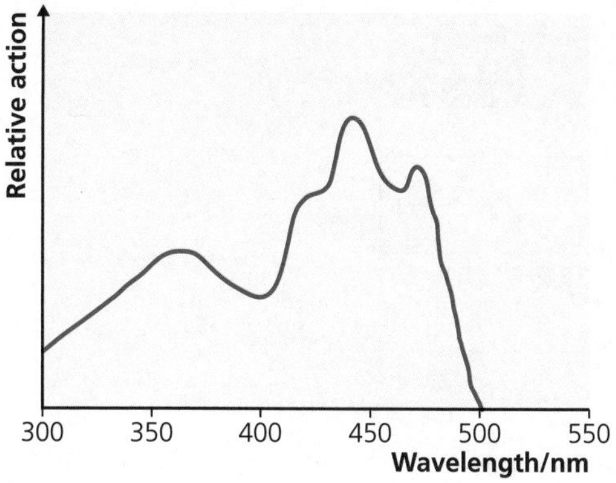

Figure 1

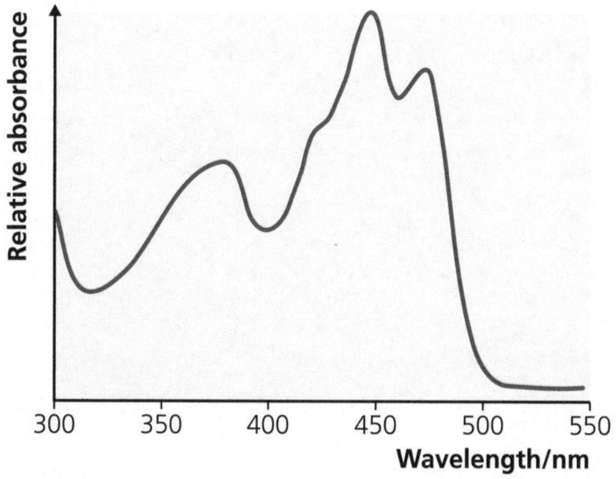

Figure 2

(a) (i) Describe the evidence shown in Figures 1 and 2 that supports the hypothesis that phototropin is the receptor in the phototropic response. (3 marks)

ⓔ You must read the introductory information and study any graphs or tables before you read the questions. This ensures that you understand it all before you focus on what you are required to do. You should annotate the text and any graphs and tables to help you.

(ii) Explain why it is necessary to be cautious when drawing conclusions based on evidence from Figures 1 and 2 alone. (2 marks)

ⓔ This is a question that tests your skills of evaluation. Look critically at the data provided and write down at least two ideas before writing an answer.

The gene **Le** codes for the enzyme 3β-hydroxylase, which catalyses the synthesis of the plant hormone gibberellin (GA_1) from its substrate GA_{20}. The recessive allele, **le**, codes for a non-functional protein.

Dwarf pea plants with the genotype **lele** were grown from seed. Several weeks later, 60 plants, all at the same developmental stage, were divided into six equal groups, A to F.

The plants in each group were watered daily with $1\,cm^3$ of a solution of gibberellic acid (GA_3). After 5 weeks the stems of each plant were measured and means calculated for each treatment. The results are shown in the table.

Group	Concentration of GA_3/$g\,cm^{-3}$	Mean length of stems/mm	Standard deviation (SD)
A	0	220	10.4
B	1×10^{-6}	176	16.8
C	1×10^{-5}	220	56.8
D	1×10^{-4}	360	137.2
E	1×10^{-3}	568	40.8
F	1×10^{-2}	536	66.4

(b) State three conditions that should be kept the same in this investigation. (3 marks)

ⓔ Make sure you do not give a condition that is given above, such as the volume of the solution.

(c) Comment on the results of this investigation, as shown in the table. (6 marks)

ⓔ 'Comment' invites you to describe, explain, criticise, evaluate and use the data to illustrate any points that you make.

Student A

(a) (i) The absorption spectrum for phototropin and the action spectrum are almost identical. For example, the maximum absorption by phototropin occurs at a wavelength of 450 nm and the greatest phototropic response occurs at the same wavelength. There must be a receptor in the plant to detect the light that comes from one side. Since phototropin absorbs strongly in the same region of the spectrum that stimulates the biggest response, it suggests that phototropin is that receptor.

e **3/3 marks awarded** This is a well-developed answer.

Student B

(a) (i) The receptors for the phototropic response are in the shoot tip. Figure 1 shows that phototropin absorbs light between 350 nm and 500 nm, with a peak at 450 nm. Figure 2 shows that the most effective wavelength for stimulating the phototropic response is also 450 nm.

e **2/3 marks awarded** Student B identifies the important evidence — the similarity between the absorption spectrum and the action spectrum — and refers to the peaks at 450 nm. However, the answer does not develop this idea further.

Student A

(a) (ii) There are two reasons to be cautious. There may be other compounds in the thale cress seedlings that could absorb light. The absorption spectrum is only for phototropin. The similarity in shape between the two graphs could just be a coincidence — phototropin could be absorbing light for another reason. Further evidence is needed.

e **2/2 marks awarded** The question is worded to prompt answers that deal with the evidence provided. Student A is correct, and further evidence is available. There are mutant alleles of the genes that code for the two types of phototropin. Seedlings of *A. thaliana* that are homozygous recessive for these genes do not show a phototropic response, as the receptor is non-functional.

Student B

(a) (ii) It is necessary to be cautious in drawing a conclusion because we are not told how often the experiment was carried out and whether there were any repeats. It could be that the results are anomalous.

e **0/2 marks awarded** Student B has given a general answer that would not gain any marks. There is enough information given in the question for a more precise response along the lines given by student A.

Student A

(b) Light intensity, temperature, humidity

e **3/3 marks awarded**

Student B

(b) Light, quantity of soil, water

e **1/3 marks awarded** 'Light' is not precise enough. Light duration (hours of light) and light intensity are acceptable answers. Quantity of soil is a good answer as the experiment is dependent on applying gibberellic acid to the soil. The volume of water given to the plants each day would be a good answer, but 'water' alone is too imprecise. Student B should realise from practical work that more precise answers are required.

Student A

(c) The dwarf plants cannot make GA_1 because the enzyme coded by the allele **le** does not function. Group A is the control — it is like the baseline. The others that were given GA_3 show greater length, apart from Group B. It could be that this is an anomalous result or that the difference between the means for A and B is not significant.

GA_3 acts to replace GA_1 and stimulate growth in height. However, we do not know whether it completely replaces GA_1 because we do not have results for a group of tall plants (**LeLe**) for comparison.

The standard deviations show that there was a wide range of stem lengths within each group, so it may not be possible to draw any conclusion about the effect of increasing the concentration of GA_3 on stem growth. They do show that it is possible to reverse the effect of the recessive allele.

e **6/6 marks awarded** This covers a range of aspects, for full marks.

Student B

(c) Plants that were watered with GA_3 grew taller than those given water alone. Group A is the control that shows some growth although we are not told how much they have grown in the 5 weeks. One group watered with GA_3 has not grown as much as group A. The highest concentration ($1 \times 10^{-2}\,g\,cm^{-3}$) gave plants that grew less than the plants in Batch E so at high concentration GA_3 is inhibiting growth.

ⓔ 1/6 marks awarded The first sentence here is incorrect. Student B is not sure how to respond to the command word *comment*. When you see this, think of different aspects. You could describe the results, look for a trend (sketching a graph is useful if the data are presented in a table) and explain the results using your knowledge. The inclusion of standard deviations suggests that you could evaluate the results, for example by giving the range of readings that includes 68% of the replicate results; in the case of D that is 222.8 to 497.2 (mean ±1SD). You could also evaluate the procedure. The point about how much the plants have grown and not knowing the stem lengths at time 0 is valid. The plants will not all have been the same height. Note that the plants in Batch F still grew a lot. At that concentration, GA_3 is not inhibiting growth.

Question 8

Figure 1 is an outline of the cycle of reactions that occurs in the light-independent stage of photosynthesis.

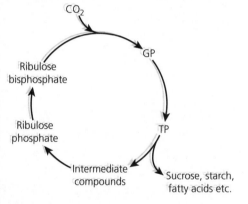

Figure 1

(a) (i) State the precise site in the chloroplast of the light-independent stage of photosynthesis. (1 mark)

ⓔ Think about your annotated revision diagram of the chloroplast before you commit yourself to an answer to this question.

(a) (ii) Show on Figure 1 where reduced NADP and ATP are used in the light-independent stage. (1 mark)

ⓔ This type of question will not have any dotted lines on the examination paper for your answer, so it is easy to miss. Always check through your paper and make sure that you have an answer for all the marks shown in brackets on the right-hand side of the page.

(a) (iii) Name the enzyme that catalyses the fixation of carbon dioxide. (1 mark)

(b) (i) Explain why the concentration of GP in chloroplasts increases as light intensity decreases at sunset. (3 marks)

ⓔ Light intensity relates to the light-dependent stage. This is the clue to unlocking this question.

(b) (ii) State and explain the effect of a decrease in carbon dioxide concentration on the concentrations of RuBP, TP and GP in chloroplasts. (4 marks)

ⓔ Use Figure 1 to help answer this question.

In 1961 the English scientist, Peter Mitchell, proposed the theory of chemiosmosis to explain how ATP is synthesised in chloroplasts and mitochondria. Figure 2 shows an experiment with grana isolated from chloroplasts that was carried out by André Jagendorf and co-workers and published in 1963.

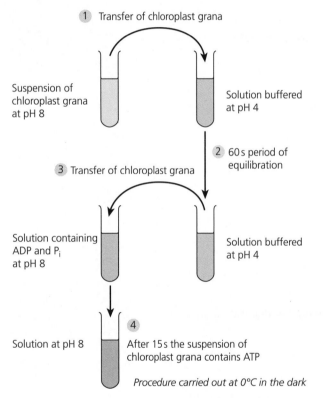

① Transfer of chloroplast grana

Suspension of chloroplast grana at pH 8

Solution buffered at pH 4

② 60 s period of equilibration

③ Transfer of chloroplast grana

Solution containing ADP and P_i at pH 8

Solution buffered at pH 4

Solution at pH 8

④ After 15 s the suspension of chloroplast grana contains ATP

Procedure carried out at 0°C in the dark

Figure 2

(c) (i) Explain why the experiment was carried out in the dark. (2 marks)

ⓔ Remember this question is about isolated grana. In the dark there is no supply of energy to drive proton pumping to create the proton-motive force (p. 69).

(c) (ii) Explain how the experiment shown in Figure 2 provides evidence for the theory of chemiosmosis. You may refer to the steps in the procedure by the numbers in Figure 2. (4 marks)

ⓔ The key word here is chemiosmosis. Think about your knowledge of chemiosmosis — membrane, energy transfer, electron flow, proton pumping, proton gradient, ATP synthase — and see how the experiment supports those ideas.

Student A

(a) (i) Stroma

ⓔ **1/1 mark awarded**

Student B

(a) (i) Stoma

ⓔ **0/1 mark awarded** Student B may well have made a slip of the pen, but unfortunately stoma is another biological term and cannot be accepted. Matrix would also be wrong as it is the term for the equivalent part in mitochondria.

Student A

(a) (ii) *Student A indicated that ATP and reduced NADP are used between GP and TP.*

ⓔ **1/1 mark awarded**

Student B

(a) (ii) *Student B indicated that reduced NADP is used between GP and TP and that ATP is used between ribulose phosphate (RP) and RuBP.*

ⓔ **1/1 mark awarded** This is also correct. Check with Figure 42 on p. 60 to confirm this.

Student A

(a) (iii) Rubisco

ⓔ **1/1 mark awarded** Rubisco is the name by which the enzyme ribulose bisphosphate carboxylase/oxygenase is universally known, so is accepted.

Student B

(a) (iii) Ribulose bisphosphatase

ⓔ **0/1 mark awarded** Student B has made an attempt at the whole name, but it is not correct.

Student A

(b) (i) As the light intensity decreases there is less energy to drive the light-dependent stage so there is less ATP and reduced NADP produced. This means that less GP is converted into triose phosphate (TP), so the concentration of GP increases and the concentration of TP decreases. Also, the reaction catalysed by rubisco (see Figure 1, RuBP → GP) continues as carbon dioxide is still available and this increases the concentration of GP. But with decreasing TP, less RuBP can be produced so the concentration of GP will reach a constant level.

ⓔ **3/3 marks awarded** Student A's answer explains the effect of light intensity as shown in Figure 43 on p. 62.

Student B

(b) (i) The concentration of GP increases because there is less photosynthesis taking place when it is getting darker at sunset. Because there is less photosynthesis there is less energy for the reactions of the Calvin cycle that fix carbon dioxide. The enzymes therefore work more slowly.

ⓔ **0/3 marks awarded** Student B states that less photosynthesis takes place when it gets darker and then states that there is less energy available. This is too general an answer. Answers should use information about ATP and reduced NADP from the light-dependent stage, especially as there is a hint about this in **aii**. This answer is not precise enough to gain any credit.

Student A

(b) (ii) When the carbon dioxide concentration decreases there is less carbon dioxide to fix. If there is less carbon dioxide, then RuBP accumulates because it is not being used to fix carbon dioxide. If carbon dioxide is not being fixed, then GP will not be formed, so the concentration decreases, as will that of TP.

ⓔ **4/4 marks awarded**

Student B

(b) (ii) When the carbon dioxide concentration decreases it becomes a limiting factor so less photosynthesis occurs and less GP is formed from RuBP. This slows down the Calvin cycle so there are fewer molecules of glucose formed:

$$6CO_2 + 6H_2O \rightarrow C_6H_{12}O_6 + 6O_2$$

ℯ 1/4 marks awarded Student B has not answered the whole of the question because there is nothing in the answer about the concentration of RuBP or TP. Again the answer is too vague to gain more than 1 mark for the reference to GP. There is no point including the equation for photosynthesis learnt for GCSE as it is not appropriate for A-level.

Student A

(c) (i) This experiment is designed to see if a proton gradient will drive the production of ATP. If this was done in the light, then energy would be available to drive the production of ATP and it would not be possible to tell what effect a proton gradient alone can have.

ℯ 2/2 marks awarded

Student B

(c) (i) If there is no light then no energy is absorbed by the pigments in the light-harvesting centres in the thylakoids. This is to see if ATP can be produced in the dark without this energy.

ℯ 0/2 marks awarded Student B gives correct information, and it is good to see energy in the answer, but does not state the reason for carrying out *this* experiment in the dark. This is about the role of the proton gradient across the thylakoid membrane.

Student A

(c) (ii) Chloroplasts have been broken apart to release the grana, which are in suspension. Each granum is made up of a stack of thylakoids. When the thylakoids are put into the solution at pH 4, protons move down their concentration gradient into the thylakoid spaces. This gives the inside of the thylakoid spaces a low pH (as they would normally have due to proton pumping, using energy from the transport of electrons from the photosystems).

The grana are transferred into a solution of pH 8, which has a low concentration of protons and is like the stroma in the intact chloroplasts. There is now a proton gradient from the thylakoid space to the stroma. ATP is formed at stage 4 in the experiment by protons diffusing through ATP synthase to form ATP from the ADP and phosphate added at stage 3.

ℯ 4/4 marks awarded This is the so-called 'acid bath' phosphorylation method that provided evidence for the chemiosmotic theory. You can expect to be asked to explain how the results of an experimental procedure support or refute a statement, theory, hypothesis or prediction. Note that the whole procedure was carried out in the dark, so there was no involvement of the photosystems and the proton pump. Student A has given a thorough answer.

Student B

(c) (ii) Grana are the site of the light-dependent stage of photosynthesis. When light strikes photosystem 2, electrons travel along the electron transport chain to photosystem 1. The energy released during the transfer of electrons is used to pump protons from the stroma into the thylakoid space. Protons can only diffuse through ATP synthase. This means that the thylakoid space has a low pH and protons diffuse through ATP synthase into the stroma down their concentration gradient and ATP is formed on the stromal side of the thylakoids, ready for the light-independent stage.

ⓔ **2/4 marks awarded** Student B has described chemiosmosis but not answered the question, scoring 2 marks for the use of appropriate information about chemiosmosis: the correct orientation of the proton gradient and the production of ATP by ATP synthase.

Question 9

The diagram shows a metabolic pathway that occurs in mammalian muscle tissue.

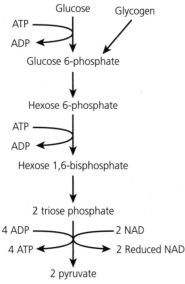

(a) (i) Name the metabolic pathway shown in the diagram. (1 mark)

ⓔ This shows the importance of making a large annotated diagram of respiration.

(ii) State the net yield of ATP when one molecule of glucose is metabolised by this pathway. (1 mark)

ⓔ Notice that this question asks for the *net* yield.

(iii) State where in muscle tissue this pathway occurs. (1 mark)

ⓔ Try to be as precise as possible when answering questions that ask 'where'.

The concentration of NAD in muscle tissue is very low — about $0.8\,\mu mol\,g^{-1}$ of muscle tissue. In aerobic respiration, reduced NAD is converted to NAD by mitochondria.

(b) (i) With reference to the diagram, explain how reduced NAD is recycled in muscle tissue when oxygen is not available. (2 marks)

ⓔ You could answer this by continuing the pathway to show how reduced NAD is recycled. Write this in the answer space. A diagram on its own will not get the marks, but it will help.

(b) (ii) Explain how reduced NAD is recycled when oxygen is available in the muscle tissue. (2 marks)

ⓔ The clue in the question is 'oxygen'. Respiration with oxygen is aerobic. That should help you to locate the part of respiration where reduced NAD is recycled and then you can explain briefly how this happens.

(c) The seeds of the castor oil plant, *Ricinus communis*, are a rich source of oils. Ricinoleic acid forms about 95% of the fatty acid content of the triglyceride molecules in these seeds. The equation for respiration of this fatty acid is:
$$C_{18}H_{34}O_3 + 25O_2 \rightarrow 18CO_2 + 17H_2O$$

(i) Use the equation to predict the respiratory quotient (RQ) of a germinating castor oil seed. Give the formula that you will use and show your working. (2 marks)

ⓔ From the information provided you have to assume that respiration will be aerobic.

(ii) An investigation found that the RQ of germinating castor oil seeds was greater than 1 at the start of germination and then fell to below 1. Suggest an explanation for this change. (1 mark)

Student A

(a) (i) Glycolysis

(ii) Two molecules of ATP per molecule of glucose

(iii) Cytosol in the muscle cells

ⓔ **3/3 marks awarded**

Student B

(a) (i) Respiration

(ii) Four

(iii) Cytoplasm

ⓔ 1/3 marks awarded Student B has confused respiration with glycolysis. The pathway shown is not the whole of respiration — with or without oxygen. Glycolysis is the metabolic pathway that is common to *both*. Cytoplasm is the site of glycolysis given in the specification, so student B gains 1 mark for part (aiii). The more precise answer to part (aiii) is cytosol. Remember that mitochondria, where the link reaction, Krebs cycle and oxidative phosphorylation occur, are also part of the cytoplasm.

Student A

(b) (i) Pyruvate acts as the hydrogen acceptor as it receives hydrogen from reduced NAD. This means that reduced NAD is oxidised and is now available for the reaction that occurs in glycolysis and is shown in the pathway. NAD is a coenzyme. When pyruvate is reduced it is converted to lactate, which diffuses out of the muscle tissue into the blood:

pyruvate + reduced NAD → lactate + NAD

ⓔ 2/2 marks awarded Student A gives a full answer. It is always a good idea to give an equation or draw part of a metabolic pathway if it helps your answer. However, do not *describe* a pathway if you are asked to explain it.

Student B

(b) (i) There is no oxygen available, so pyruvate does not enter mitochondria to be respired aerobically. Instead, it is converted to lactate, which leaves the muscle.

ⓔ 1/2 marks awarded Student B gains 1 mark for stating that pyruvate is converted to lactate. There is nothing in the answer about NAD.

Student A

(b) (ii) When oxygen is available, reduced NAD is recycled in mitochondria. It is oxidised by intermediate compounds, which pass into the matrix of the mitochondrion. The hydrogens from reduced NAD from glycolysis are passed to the ETC.

ⓔ 2/2 marks awarded Student A makes it clear that reduced NAD is recycled by the action of mitochondria. It is not necessary to know the details of the recycling of NAD from glycolysis (such as the names of the intermediate compounds), but you should know that hydrogen ions and electrons from reduced NAD are made available to the electron transport chain (ETC). Student A uses an accepted abbreviation here.

Student B

(b) (ii) Oxygen is the final electron acceptor in aerobic respiration. This is how reduced NAD is recycled.

ⓔ 0/2 marks awarded Student B has jumped to the end of the story of oxidative phosphorylation, so gains no marks.

Student A

(c) (i) $RQ = \dfrac{\text{volume of carbon dioxide produced}}{\text{volume of oxygen absorbed}}$

$= \dfrac{18}{25} = 0.72$

ⓔ 2/2 marks awarded

Student B

(c) (i) RQ = 1.39

ⓔ 0/2 marks awarded Student B has not answered the question by giving the formula and has not shown any working. The answer shows that the student has forgotten the formula and used it the wrong way round.

Student A

(c) (ii) Respiration at the start of germination was anaerobic; then oxygen diffused throughout the seed so that respiration became aerobic.

ⓔ 1/1 mark awarded

Student B

(c) (ii) To begin with the seed respired very slowly and then the seed coat broke open and it respired much faster.

ⓔ 0/1 mark awarded RQ does not tell us anything about the rate of respiration, only about the type of respiration and the type of substrate.

ⓔ Overall, student B gains 17 marks out of 55. This sort of performance will not be good enough for a pass grade on Paper 1 as a whole. Marks were lost for a number of reasons:

- Using information from GCSE that is not of the depth required at A-level (e.g. Q.8bii).
- Not developing answers fully (e.g. Q.7ai, Q.8bi).
- Not using terms from the specification (e.g. Q.6b).
- Describing information, rather than explaining it (e.g. Q.8b, Q.8cii).
- Not understanding what is required for an answer (e.g. Q.6b, Q.8ci, Q.9ai).
- Not following the command word carefully (e.g. Q.7aii and Q.7c).
- Not answering precisely enough (e.g. Q.7b and Q.8aiii).
- Not answering in sufficient detail (e.g. Q9bii).
- Not learning formulae (e.g. Q.9ci).
- Not answering the question in terms of the data supplied (e.g. Q.8cii).
- Not answering the question (e.g. Q.9cii).

■ Paper 3-style questions: Unified biology

Question 1

The black desert grasshopper, *Taeniopoda eques*, is an ectotherm. These grasshoppers orientate themselves in different directions to the Sun during the day. Sometimes they expose the sides of their bodies (flanking position) and at others they turn facing the direction of the Sun (head facing position). Two of these grasshoppers were kept restrained in these positions and their body temperatures monitored over 25 minutes. The results are shown in the diagram.

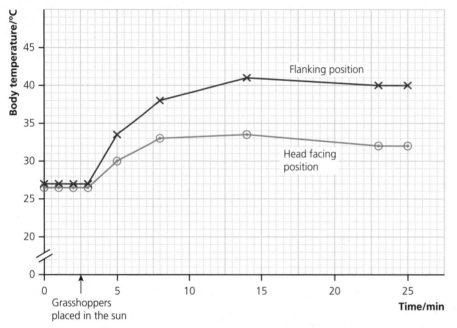

(a) Describe and explain the results. (5 marks)

ⓔ Write side headings in the space provided for your answer so you remember to give a description with figures (including units) and an explanation.

(b) Black desert grasshoppers also adopt other positions, such as pressed down on the ground and sitting on the branches of trees. Suggest how these positions help the grasshoppers to regulate their body temperature. (2 marks)

ⓔ The clue here is in the last four words — the question does not say 'raise' or 'maintain' their body temperature.

(c) Sand dune lizards appear to be more active at higher temperatures. Outline how you would investigate the effect of ambient temperature on the activity of sand dune lizards. (6 marks)

ⓔ This question tests your skills of planning. Always state how you would collect and analyse results.

Student answers

(a) Both grasshoppers show an increase in body temperature for the first 5 minutes of exposure. The grasshopper with flank facing reaches a higher body temperature. The body temperature of the grasshopper with head facing the Sun increases from 27°C to 32°C; the body temperature of grasshopper with flank facing the Sun increases from 27°C to 41°C. Both grasshoppers absorb heat by radiation; flanking exposes a greater surface area to the Sun.

e **5/5 marks awarded**

(b) If the body temperature is colder than the surroundings grasshoppers can absorb heat by conduction from the soil. If the body temperature is warmer than the surroundings then they can lose heat by conduction to the soil. When sitting in branches they can lose heat by convection.

e **2/2 marks awarded** For (a) and (b), it helps to apply your knowledge of methods of heat transfer by radiation, conduction and convection.

(c) 1 Decide on a way to record the activity of sand lizards, for example by using direct observation, videoing or time-lapse photography.

2 Record the activity of one or more lizards at intervals of time during 24-hour periods.

3 Classify the behaviour of the lizards, for example basking, hunting, interacting with other lizards, resting.

4 Record the air and ground temperatures throughout the investigation using sensors and a data logger.

5 If using a lab investigation animals should be kept in an environment that resembles their natural habitat (not an empty cage).

6 In the lab, the temperature can be changed by using infrared lamps and a thermostat.

7 Analyse results to see which is the most common activity at different temperatures.

8 Express results as the percentage of the time resting and being active.

e **6/6 marks awarded** Procedures are often easier to set out as numbered points.

Question 2

Figure 1 is a false-colour transmission electron micrograph showing a junction between two neurones in the spinal cord.

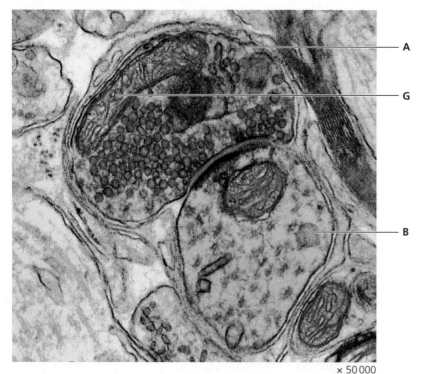

× 50 000

Figure 1

(a) (i) Calculate the maximum length of the organelle labelled **G**. Give the formula that you will use and show your working. 🖩 (2 marks)

ⓔ Remember to measure in mm, not in cm. Give your answer in micrometres or use standard form if you use mm.

(ii) Identify organelle **G** and give a reason for your answer. (2 marks)

ⓔ Some organelles have a habit of not looking like textbook examples. This is one example.

(b) Identify neurones **A** and **B** and relate their appearance to their functions. (5 marks)

ⓔ If you are unsure about a question, read the stem again. The clue here is in the sentence above the TEM.

The concentration of calcium ions in the cytosol of all cells, including neurones, is very low. Calcium ions enter axon terminals on the arrival of impulses.

(c) Suggest how the concentration of calcium ions in the cytosol of axon terminals is maintained at a very low level. (3 marks)

ⓔ 'Suggest' implies that you are not expected to know the answer, but can work it out by applying knowledge of other topics. Here you can use the resting potential of neurones and the role of sarcoplasmic reticulum in muscle fibres.

(d) Explain the role of calcium ions at axon terminals. (3 marks)

ⓔ Making a large table to record information about the ions that are listed in Module 2 (including calcium) will help with answering questions like this one.

Pacinian corpuscles and Ruffini endings in the skin are mechanoreceptors. The responses of these two receptors to a single, sustained stimulus were recorded as shown in Figure 2.

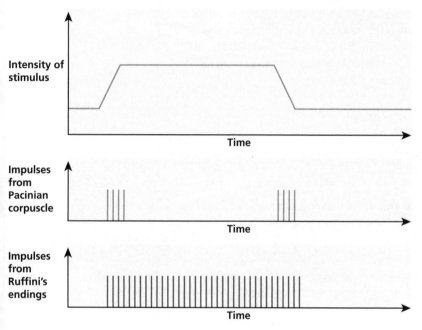

Figure 2

(e) Compare the activities of the two skin receptors in response to the sustained stimulus. (3 marks)

ⓔ Compare means that you should give similarities as well as differences.

Student answers

(a) (i) actual length = length on image/magnification

= 50 mm/50 000 = 1.0 × 10⁻³ mm = 1.0 µm

ⓔ **2/2 marks awarded**

(a) (ii) Mitochondrion. Double membrane; inner membrane is folded to form cristae.

ⓔ 2/2 marks awarded Make sure you look at plenty of TEMs of all the specialised cells listed in the specification.

(b) The small circles coloured red in neurone **A** are vesicles containing neurotransmitter, so **A** is the pre-synaptic neurone and **B** is the post-synaptic neurone. **B** does not have vesicles, so impulses can only travel in one direction — from **A** to **B**.

A could be a sensory neurone or a relay neurone. **B** could be a relay neurone or a motor neurone.

When **A** transmits impulses across the synapse the vesicles move towards the membrane and release molecules of neurotransmitter by exocytosis. The post-synaptic neurone (either relay neurone or motor neurone) will have receptors for the neurotransmitter. These will be stimulated by the neurotransmitter and if the resultant depolarisation is above the threshold, then the post-synaptic neurone will send impulses.

ⓔ 5/5 marks awarded This is a thorough answer.

(c) Carrier proteins in the membrane pump calcium ions out of the neurone by active transport, using ATP from mitochondria.

ⓔ 3/3 marks awarded An alternative answer is that mitochondria act like the sarcoplasmic reticulum in muscle fibres to concentrate (sequester) calcium ions.

(d) Depolarisation of the membrane stimulates voltage-gated calcium ion channel proteins to open. Calcium ions enter by facilitated diffusion and act as a second messenger, stimulating movement of vesicles to the pre-synaptic membrane.

ⓔ 3/3 marks awarded This is one of several roles of calcium ions that you can learn for Module 2.

(e) Both receptors start sending impulses when the stimulus is applied. The Pacinian corpuscle sends impulses when the stimulus starts and ends but not when it is applied constantly; Ruffini endings send impulses at a constant rate while the stimulus is applied.

ⓔ 3/3 marks awarded The marks are only awarded when direct comparisons are made — they should include similarities and differences.

Question 3

Kenneth Thimann and Folke Skoog published a scientific paper in 1933 providing evidence for the role of auxin in controlling apical dominance. Some researchers have found it difficult to repeat their results with other species in different environmental conditions. In the mid-1990s, Morris Cline repeated these experiments with several species and concluded that Japanese morning glory, *Ipomoea nil*, gave the most repeatable results.

Plants of *I. nil* were grown from seed until they had several nodes with lateral buds. The apical buds were cut from some of the plants and they were divided randomly into three batches, with the shoot stumps covered with lanolin cream:

A without any added auxin

B with 0.01% auxin solution

C with 0.1% auxin solution

A fourth batch (D) was included that did not have apical buds removed.

The plants were grown in identical conditions. The lengths of the lateral shoots on each plant were measured over 4 days. The results are shown in the table.

Time after decapitation/days	Mean length of lateral shoots/mm ± SD			
	A	B	C	D
0	2 ± 1	3 ± 0	3 ± 1	1 ± 0
1	3 ± 1	3 ± 1	3 ± 1	1 ± 0
2	6 ± 1	3 ± 1	3 ± 1	1 ± 0
3	12 ± 3	6 ± 1	5 ± 1	1 ± 0
4	23 ± 6	11 ± 3	8 ± 2	2 ± 1

(a) Plot a graph showing changes in mean length of lateral shoots in batches A, B and C.

(5 marks)

ⓔ A grid would be printed on the exam paper that would be large enough for you to scale the axes so that they make full use of the space provided.

(b) Explain how the results support the hypothesis that auxin controls apical dominance.

(4 marks)

ⓔ Although only A, B and C are to be plotted on the graph, refer to D as well. What does this control tell you?

(c) Researchers in 2014 provided evidence for an alternative mechanism for apical dominance, since it had been discovered that auxin does not flow from apical buds into lateral buds. Their results showed that:
- lateral buds start to grow before auxin concentrations decrease in the stem
- the sucrose concentration in lateral buds increases immediately the apical buds are removed
- the increase in sucrose leads to the repression of the gene *BRC1*, which codes for a protein that inhibits the transcription of genes involved in growth of apical buds

> **(i)** Explain how the concentration of sucrose in lateral buds might increase following removal of the apical bud. (3 marks)

ⓔ Thorough knowledge of photosynthesis and translocation (from Module 2) is essential for this question.

> **(ii)** Suggest how sucrose might act to repress the gene *BRC1*. (2 marks)

ⓔ This is set in the context of gene control from Module 6.

Student answers

(a) *Time is the x-axis and mean length is the y-axis. The axes should be scaled using the full space on the grid. The axes should have titles taken from the table. The question asks for the mean results, so SD error bars are not required. The points should be plotted clearly and joined with straight lines. Each line must be labelled (A to C).*

(b) The apical bud is the main source of auxin, which travels down the stem and inhibits the growth of lateral buds, so the plant grows in height. Without the apical bud the lateral buds begin to grow. The application of auxin to the cut stumps restored the supply of auxin and in B and C there was less growth in lateral buds — for example, by day 4 the growth of lateral buds in C was a third that of A. There was less growth in the higher concentration (C) than in the lower concentration (B), showing a dose response.

ⓔ Always quote from the data, which can be derived from the table or graph, as here. Ratios, proportions and percentages are all appropriate.

(c) (i) Sucrose travels from the leaves (source) to sinks, such as buds and roots. When the apical bud is removed the lateral buds now become sinks so phloem transports sucrose from leaves to lateral buds instead of to the apical bud.

ⓔ Always use correct terminology if you can. Here source, sink and phloem are good terms to use. Do not just refer vaguely to transport in the stems.

(i) Transcription of genes in eukaryotes requires a variety of transcription factors, which are proteins. There will be several transcription factors for *BRC1*. Sucrose could combine with one of these transcription factors to prevent it helping to activate transcription by RNA polymerase. The relevant transcription factor will have a sucrose-binding region and will change shape when sucrose binds.

ⓔ From Module 6 you will know about the role of transcription factors in pre-transcriptional regulation of gene expression in eukaryotes. You will also know about this type of regulation in prokaryotes, which is much simpler.

Answers to multiple-choice answers

Question	Answer
1	C
2	B
3	C
4	D
5	D

Knowledge check answers

1 Ectotherms gain heat from their surroundings; endotherms generate heat in their bodies, for example through respiration and shivering.

2 Hypothalamus.

3 It reduces heat loss by physiological and behavioural means. A mammal conserves heat by contracting hair erector muscles to give a greater depth of fur, reducing blood flow through the outer capillaries of the skin, and curling up to reduce surface area exposed to the cold air. It will also generate heat by shivering and increasing the rate of metabolism in the liver. Some mammals have brown fat, which generates heat.

4 Negative feedback acts to maintain a factor at a near constant level; positive feedback acts to continually increase the factor. Negative feedback occurs continually; positive feedback is a short-lived mechanism.

5 18.9%/19%. The volume of plasma that passes through the glomeruli every minute is 660 cm^3.

6 Glucose, ions (e.g. sodium and chloride), urea and water are reabsorbed; the volume of filtrate decreases significantly. Reabsorption occurs by active transport (glucose and sodium ions), by diffusion (urea) and by osmosis (water).

7 The pore in the centre of the aquaporin is just large enough for water but does not let larger molecules or ions through; the channel is lined by amino acid residues that are charged and repel ions of the same charge.

8 The relative molecular mass of insulin is below 69 000, so it is filtered through the glomerulus and is not reabsorbed.

9 Receptors convert the form of energy of a specific stimulus into electrical impulses in neurones.

10 Three factors influence the speed of conduction of neurones. Myelinated neurones are much faster than unmyelinated neurones. Conduction along myelinated neurones is faster because the impulse jumps from node to node. Wide neurones are faster than narrow neurones. Resistance decreases as the cross-sectional area increases (you probably know this from your knowledge of physics). Speed of conduction is also influenced by body temperature; it is faster in animals with high body temperature, as the movement of ions through channel proteins and their opening and closing is temperature dependent.

11 Sensory neurones transmit impulses from receptors to the central nervous system (CNS); relay neurones transmit impulses between sensory and motor neurones; motor neurones transmit impulses from the CNS to effectors (muscles and glands). Avoid using the terms 'message' and 'signal' when writing about neurones.

12 110 mV; this varies between neurones depending on the resting potential and the maximum potential difference at the height of the action potential.

13 There are several answers to this:
- Information goes from receptor to CNS or from CNS to effector; there is no point in information going in the reverse directions as receptors receive stimuli and effectors make the changes required.
- Synapses are polarised — with vesicles of neurotransmitter on one neurone and receptors for these cell-to-cell signalling compounds on the other, so nerve impulses go in one direction.
- The region of a neurone behind an action potential is undergoing the refractory period; voltage-gated sodium ion channel proteins are inactivated so will not open, so the impulse cannot go backwards to where it has just come from.

14

Stage of action potential	Voltage-gated channel proteins	
	Sodium	Potassium
During depolarisation towards threshold	Activation gates on some channel proteins opening	All activation gates closed
Rising phase	More activation gates opening	Most activation gates closed; some opening slowly
Falling phase	Inactivation gates closing quickly	Activation gates opening slowly
Refractory period	All inactivation gates closed	All activation gates open

15 Depolarisation stimulates voltage-gated calcium ion channel proteins to open, allowing calcium ions to flow into the synaptic bulb; this stimulates vesicles to move towards the presynaptic membrane; vesicles fuse with the membrane, releasing neurotransmitter molecules into the synaptic cleft (exocytosis); neurotransmitter molecules diffuse across the synapse and bind with ligand-gated sodium ion channel proteins; these channel proteins open to allow sodium ions to diffuse into the postsynaptic neurone.

16 Exocrine — secretion into a duct — substance(s) secreted travel down a duct to outside (sweat duct) or into another organ (e.g. from liver and pancreas into duodenum); endocrine — secretion into the blood (not into a duct).

17 Insulin stimulates the uptake of glucose from the blood and its storage as glycogen; glucagon stimulates the breakdown of glycogen to produce glucose, which diffuses into the blood.

18 This table has some points of comparison; you may be able to think of others.

Feature	Glycogen	Glucagon
Type of biochemical	Carbohydrate Polysaccharide	Protein Single polypeptide
Molecular structure	Branched	Unbranched
Monomer	α-glucose	Amino acids
Function	Energy storage	Signalling molecule (hormone)
Site of production	Liver cells, muscle cells	α-cells in the islets of Langerhans in the pancreas

19 Type 1 diabetes is caused by an inability to produce insulin; type 2 diabetes is an inability of target cells to respond to insulin.

20 Stem cells could be placed into the body; they differentiate into islet β-cells and secrete insulin. To protect against destruction by the immune system they could be encapsulated before transplant.

21 Positive chemotropism

22 The effects are counteractive, for example the effects of glucagon and insulin.

23 ATP from aerobic and anaerobic respiration and from creatine phosphate

24 Mitochondria produce carbon dioxide and water, which are raw materials for photosynthesis in chloroplasts. The products of photosynthesis are triose phosphate (which can be converted to pyruvate in glycolysis) and oxygen (used by mitochondria).

25

Feature	Cyclic photophosphorylation	Non-cyclic photophosphorylation
Photosystem	I	I and II
Photolysis	No	Yes
Electron donor	P700 in photosystem I	Water
Last electron acceptor	P700 in photosystem I	NADP
Products	ATP	ATP; reduced NADP; oxygen

26 X — light intensity; Y — carbon dioxide concentration

27 Extra lighting to increase light intensity; heaters to increase temperature; ventilation to decrease temperature (too hot and enzymes denature); carbon dioxide enrichment by burning gas or releasing carbon dioxide (by-product of fermentation industries).

28 Cytosol, inner mitochondrial membrane, thylakoid membrane

29 Malonate is a competitive inhibitor of the enzyme. You studied enzyme inhibitors in Module 2 (covered in the first student guide of this series). There will be many questions on topics in Modules 2, 3 and 4 in the A-level papers.

30 Glycolysis — cytosol; link reaction and Krebs cycle — matrix of mitochondrion; oxidative phosphorylation — inner mitochondrial membrane

31 NAD and FAD are hydrogen carriers; co-enzyme A transfers two carbon acetyl groups into the Krebs cycle.

32 Lactate is an energy-rich molecule; excreting it would be a waste of energy; cardiac muscle respires lactate; it can be converted into glycogen and stored for future use.

33 Lipids are energy-rich because they are highly reduced, with a larger proportion of hydrogens to carbon than carbohydrates. On oxidation during respiration they release more energy, as more reduced hydrogen carriers are produced per mole.

Index